ADVANCED LONG RANGE SHOOTING

Learn vital shooting skills from professionals who instruct foreign snipers.

Tell people what you think? Thanks!

Reviews from amazing people like you help other marksmen discover these insights, and make the entire community better and smarter.

Thank you in advance for your help and time!

Become one of the most precise and accurate marksmen on the planet.

Multivariable Shooting

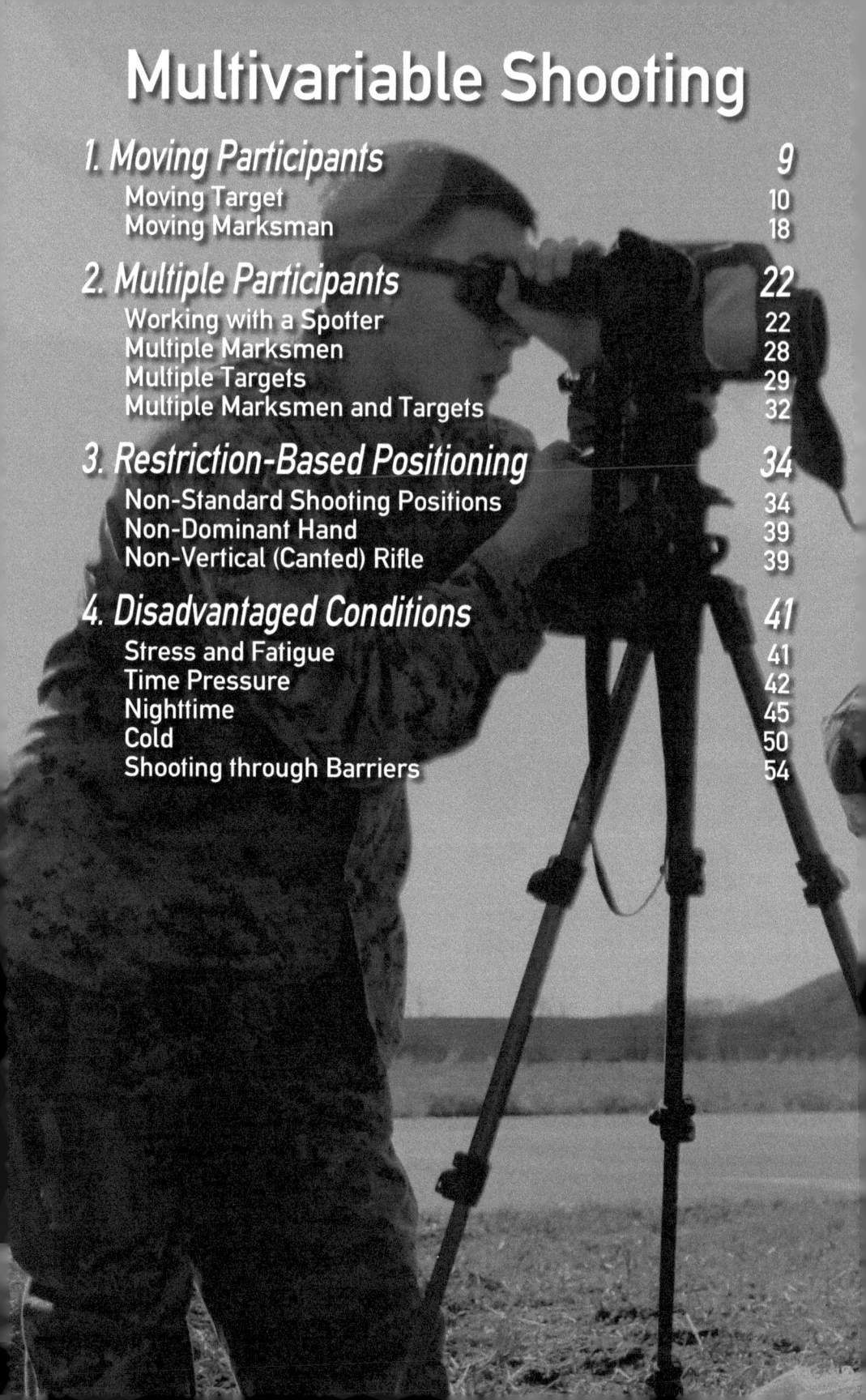

Advanced Effects, 600 Meters and Beyond

De Minimis Effects

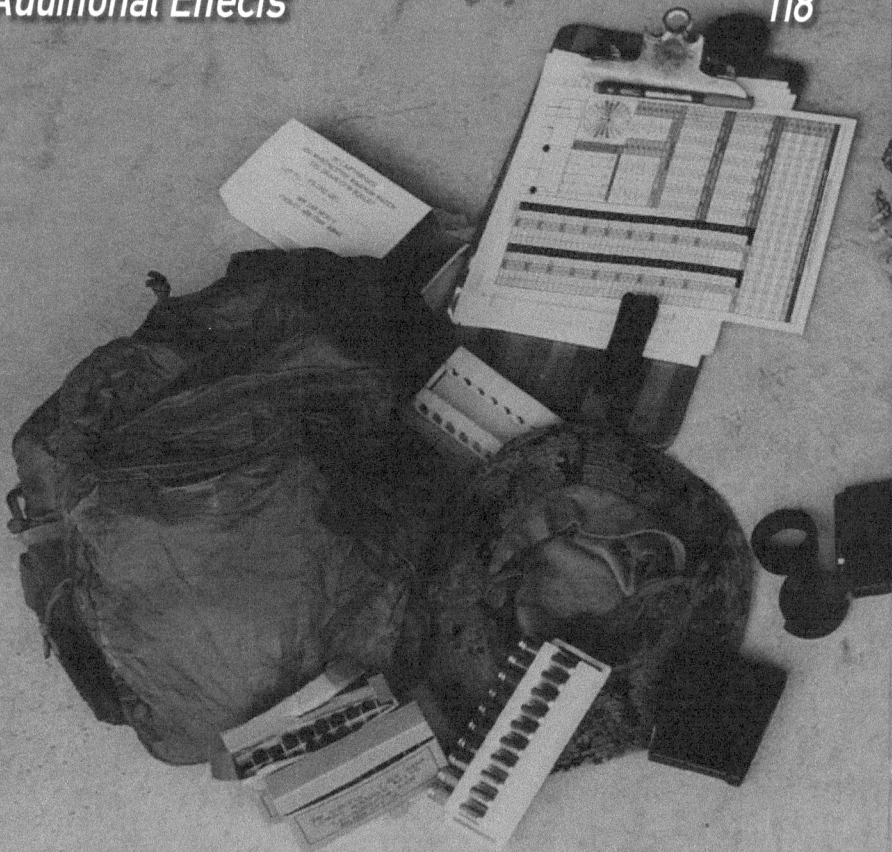

Appendices

Multivariable Shooting

Multivariable Shooting

Practice.
—Simo Häyhä, the deadliest sniper in history with over 500 confirmed kills, when asked how he became such a good marksman.

Almost all shooting is performed as a single shot from a stationary marksman to a (relatively) stationary target under daylight conditions. Normally, the only two major considerations (or variables) a marksman must consider are the distance to the target and the wind.

However, when these **marksmen incorporate additional factors**, they are engaging in multivariable shooting. Such additional factors may include: how others act and react, movement of themselves and targets, restricted positions, and abnormally bad conditions.

Some hobbyists seek the challenge of multivariable shooting, while hunters often want to expand their skills and available targets. In contrast, occupational marksmen (i.e., military and police) sometimes have no choice but to take multivariable shots because they must complete a mission.

1. Moving Participants

Sometimes, a marksman is aiming at a moving target. Sometimes, the marksmen themselves are moving. And in rare instances, both the target and the marksman are moving.

In any case, accounting for movement is related to two techniques already explained in *Long Range Shooting: An Illustrated Manual*: **elevation and windage**. "Elevation" is the vertical adjustment a marksman must make with their rifle, typically measured using the reticle, to ensure bullets reach distant targets. When a target moves closer or farther away, it changes the distance that a bullet must travel. To compensate for the changed distance, the marksman compensates by adjusting the elevation setting higher or lower to alter the trajectories of their bullets.

"Windage" is the left-or-right distance that a marksman must move their rifle to account for wind pushing the bullet sideways (and any other sideways consideration). If movement causes the target to be to the side of the point-of-aim, the marksman uses a windage hold or dial just as if wind pushed the bullet to the side. In fact, when a target is moving in the wind, both windage holds are simply added together.

1.a Moving Target

A moving target can be a deer in the woods sauntering around, an enemy patrolling, or a target on a moving rail system. When considering moving targets, the first consideration is simply to **wait for a target to stop moving**. Although it seems obvious, many marksmen get caught up in the moment and engage too quickly. In fact, most hunters deliberately avoid shooting at moving targets because they consider maiming an animal to be unethical and are unwilling to accept the slight decrease in accuracy that comes with shooting a moving target. Therefore, these hunters wait for a deer to stop or slow down, maximizing their chances of achieving a quick and accurate kill. **A full stop is not necessary**, and simply waiting for the target to slow down or change direction can be enough to greatly increase accuracy.

If the target does not slow down, the second thing that a marksman must determine is how long their bullet's flight time would be. Bullets take time to go from the rifle to the target, and during that time the target moves. Therefore, by determining their bullet's flight time, the marksman is also considering the target's movement time. In other words, if the marksman places the point-of-aim directly on the target, the target would move out of the way as the bullet flies. Therefore, the marksman must fire while their point-of-aim is in front of the target so that both the target and the bullet simultaneously intersect their movements at the point-of-aim.

This sideways distance that the target moves while the bullet is airborne is called the "**lead**" (rhymes with "seed"). To determine the lead, a marksman must know the sideways speed of the target and their bullet's flight time (i.e., the amount of time that the target has to move). For example, if a 400-m-or-yd target is trotting at a pace of 0.5 m or yd/s and a bullet takes 2 s to travel 0.5 m or yd to a target, the marksman must aim 1 m or yd ($0.5 \times 2 = 1$) in front of the target for the bullet and the target to intersect at the point-of-aim.

A bullet's flight time can be calculated using two different methods. The simplest method is to **fire a bullet and time it**. If a marksman does this at various distances, they can prepare a chart that matches target distances to flight times. These charts can record the flight time as a function of distance (e.g., 0.45 s at 500 m). Many of these tables are available online or preprogrammed into ballistic calculators. Then, once the actual target's distance is known, the marksman can look up that target distance in their table to find the corresponding bullet flight time.

To create these tables oneself, a marksman can go to a long-distance range with a recording device. They shoot at metal targets spaced 100 m or yd

Image 1: Marines set and prepare Robotic Moving Targets to shoot at. Quantico, VA, 24 Sep 2013.

apart. Then, at home, they determine the travel time by measuring the time difference in the audio recording between the shot and the impact. If the target is too far to hear, the same process can be used with a video recording. A little bit of extra time must be added to the final number to account for the time that it takes for a bullet to exit the muzzle after the trigger is pulled.

The second method uses calculations to determine the travel time. That is, a marksman **multiplies the average velocity of the bullet by the distance** that the bullet travels. The simplest way to "calculate" average velocity is to substitute in muzzle velocity. This method works over short distances since velocity does not change much at the start of its trajectory. For example, if the muzzle velocity of a bullet is 800 m/s, then the marksman can be relatively certain that a bullet would take about ½ s to travel 300 to 400 m.

Once a target is far away, bullets slow down significantly. At that point, substitution no longer works. Instead, a marksman can calculate the average velocity by using a ballistic calculator. A good ballistic calculator can give the bullet velocity at various points in the trajectory. By averaging together a few points, the marksman can get a good approximation of the average velocity. For example, although not exact, a marksman can then take a mean of: the muzzle velocity, the mid-distance velocity, and the velocity at the target.

Once the bullet's flight time is calculated, the marksman must determine the target's **lateral speed**. That is, if a target is moving diagonally, only the

Movement Value Chart

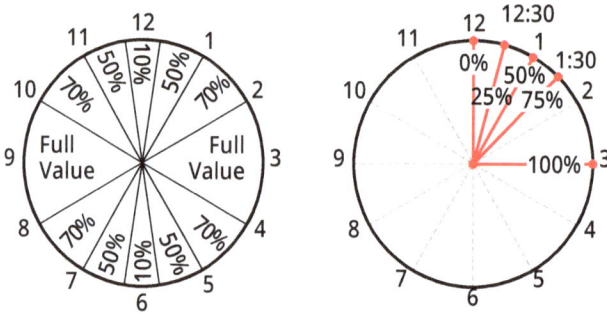

Image 2: **Movement value** is based on where the target is moving to. (Movement values and wind values are identical since they are both based on the same trigonometry. However, instead of the wind pushing the bullet back to the target, the target moves into the path of the bullet.)

The simplest movement value system is the **clock system**, where the marksman points at 12 o'clock. A target moving parallel has no value, while one moving sideways has a full value. On the left is a full system with only 4 values: 100%, 70%, 50%, and 10%. This system becomes less precise as the target strays from the 12 or 6 o'clock, because the percentages change quickly there. A filled in table can be useful for showing more numbers. (See Image 3, Pg. 13.)

---

side-to-side component is useful since a target is unlikely to outrun a bullet when moving closer or farther from the marksman. The only possible exception to that may be vehicles where it may be required to determine their front-back speed. In that case though, the left-right and front-back components of movement are still separated because they are separately accounted for with windage and elevation respectively. (In fact, a shot at a moving helicopter would, in theory, additionally require determining vertical velocity.)

Lateral velocity can be measured with a reticle's windage hashmarks. For example, a marksman can look through their scope and measure how far across the reticle the target traveled in 1 s. Or the marksman can determine how long it took the target to traverse 0.5 mils. Either method measures the angular distance the target traveled per unit of time.

Getting the angular velocity is very helpful because it can be multiplied by the bullet's flight time to directly calculate the windage hold. For example, if a target is moving right-to-left at 0.5 mil/s, and the flight time is 0.2 s, then the lead is 0.1 mils (0.5 mil/s × 0.2 s = 0.1 mils).

The other methods for getting the target's angular velocity require knowing the linear velocity and converting it to an angular velocity. The

Lateral Movement Value

Target Direction		Target Speed m/s (ft/s)									
Degrees	O'clock	1 (3.3)	2 (6.6)	3 (9.8)	4 (13)	5 (16)	6 (20)	7 (23)	8 (26)	9 (30)	10 (33)
0°	12:00, 6:00	0 (0)	0 (0)	0 (0)	0 (0)	0 (0)	0 (0)	0 (0)	0 (0)	0 (0)	0 (0)
10°		0.2 (0.7)	0.3 (1)	0.5 (1.6)	0.7 (2.3)	0.9 (3)	1 (3.3)	1.2 (3.9)	1.4 (4.6)	1.6 (5.2)	1.7 (5.6)
15°	11:30, 12:30, 5:30, 6:30	0.3 (1)	0.5 (1.6)	0.8 (2.6)	1 (3.3)	1.3 (4.3)	1.6 (5.2)	1.8 (5.9)	2.1 (6.9)	2.3 (7.5)	2.6 (8.5)
20°		0.3 (1)	0.7 (2.3)	1 (3.3)	1.4 (4.6)	1.7 (5.6)	2.1 (6.9)	2.4 (7.9)	2.7 (8.9)	3.1 (10)	3.4 (11)
30°	11:00, 1:00, 5:00, 7:00	0.5 (1.6)	1 (3.3)	1.5 (4.9)	2 (6.6)	2.5 (8.2)	3 (9.8)	3.5 (12)	4 (13)	4.5 (15)	5 (16)
40°		0.6 (2)	1.3 (4.3)	1.9 (6.2)	2.6 (8.5)	3.2 (11)	3.9 (13)	4.5 (15)	5.1 (17)	5.8 (19)	6.4 (21)
45°	10:30, 1:30, 4:30, 7:30	0.7 (2.3)	1.4 (4.6)	2.1 (6.9)	2.8 (9.2)	3.5 (12)	4.2 (14)	4.9 (16)	5.7 (19)	6.4 (21)	7.1 (23)
50°		0.8 (2.6)	1.5 (4.9)	2.3 (7.5)	3.1 (10)	3.8 (13)	4.6 (15)	5.4 (18)	6.1 (20)	6.9 (23)	7.7 (25)
60°	10:00, 2:00, 4:00, 8:00	0.9 (3)	1.7 (5.6)	2.6 (8.5)	3.5 (12)	4.3 (14)	5.2 (17)	6.1 (20)	6.9 (23)	7.8 (26)	8.7 (29)
70°		0.9 (3)	1.9 (6.2)	2.8 (9.2)	3.8 (13)	4.7 (15)	5.6 (18)	6.6 (22)	7.5 (25)	8.5 (28)	9.4 (31)
75°	2:30, 3:30, 8:30, 9:30	1 (3.3)	1.9 (6.2)	2.9 (9.5)	3.9 (13)	4.8 (16)	5.8 (19)	6.8 (22)	7.7 (25)	8.7 (29)	9.7 (32)
80°		1 (3.3)	2 (6.6)	3 (9.8)	3.9 (13)	4.9 (16)	5.9 (19)	6.9 (23)	7.9 (26)	8.9 (29)	9.8 (32)
90°	3:00, 9:00	1 (3.3)	2 (6.6)	3 (9.8)	4 (13)	5 (16)	6 (20)	7 (23)	8 (26)	9 (30)	10 (33)

Image 3: This table takes the target's actual movement speed (X-axis) and the target's direction (Y-axis), and outputs the **lateral movement velocity (cell contents)**. The value is found by multiplying the actual movement velocity by the sine of the degree when the direction the marksman faces is defined to be 0°. This is an expanded version of the above image.

linear velocity of a target can be estimated from prior knowledge, or it can be measured precisely in the moment. An example of an estimate based on prior knowledge is that people typically walk at about 1 m/s (3.3 ft/s), run at 2 m/s (6.6 ft/s), and sprint at 4 m/s (13 ft/s). (The average individual is often in poor shape.) A healthy adult male can sprint at 7 m/s (23 ft/s), while Usain Bolt set a world record of 12.42 m/s (40.75 ft/s).) An experienced marksman can refine these numbers by differentiating between different gaits and different groups of people. For example, a man may jog at 1.5 m/s (4.9 ft/s) while a woman may typically be 0.25 m/s (0.8 ft/s) slower than a man due to their shorter stature; therefore, women may jog at 1.25 m/s (4.1 ft/s).

Speed can also be estimated based on body width. A healthy man has a chest depth of about ¼ to ⅓ m (10 to 13 in), so if they travel 6 body-widths per second, they are traveling at 2 m/s (6.6 ft/s). Body-widths can even be a measurement in and of themselves. For example, if the marksman knows that a bullet would take 1 s to reach the target, they can aim 6 body-widths ahead.

Finally for this method, linear, lateral velocity is converted into angular velocity. The formulas for mils and MOA to **calculate an object's angular velocity** *(T Angular V)* with a known distance to target *(D to T)* and an estimated linear velocity in m or yd *(T Linear V)* are:

▸   *(T Angular V in Mils) = ((T Linear V) × 1000) ÷ (D to T)*
▸   *(T Angular V in MOA) = ((T Linear V) × 3438) ÷ (D to T)*

Once a marksman knows the target's actual velocity, they must then **derive the component of that velocity that is perpendicular** (i.e., lateral) to the trajectory of the bullet. The simplest method of deriving lateral velocity from actual velocity is to place the target into one of three categories: full turn (100% of actual velocity), half turn (70% of actual velocity), and no turn (0% of actual velocity). (See Image 2, Pg. 12.)

How far a man or animal is turned can be estimated by the number of arms or legs that are visible. That is, if only one arm is visible then they have a full turn; if both arms are visible but one shoulder is a little hidden, then they have a half turn; and if both shoulders are clearly visible then the target has no turn. Of course, a marksman can create as many categories and corresponding lead multipliers as·they can reasonably employ.

Putting this together with an example, assume that a target at 350 m is walking diagonally at 1 m/s. Because they are moving diagonally (i.e., half turn), the turn multiplier is 70%, so they have 0.7 m/s of sideways movement. Using the above formula to find the target's angular, lateral velocity:

▸   *(T Angular V in Mils) = ((T Linear V) × 1000) ÷ (D to T)*
▸   *(T Angular V in Mils) = ((0.7 m/s) × 1000) ÷ (350 m) =* **2 mils/s**

The resulting angular velocity is then multiplied by the bullet's flight time to determine the lead. Assume a bullet has an **average** velocity of 1000 m/s. The formula for flight time is:

▸   *(Bullet Time) = (D to T) ÷ (Bullet V)*
▸   *(Bullet Time) = 350 m ÷ 1000 m/s =* **0.35 s**

The final windage hold is determined by this formula:

▸   *(Windage in Mils) = (T Angular V in Mils) × (Bullet Time)*
▸   *(Windage in Mils) = 2 mils/s × 0.35 s =* **0.7 mils**

Movement to Windage in Mils (MOA)

Target Lateral Velocity m/s (ft/s)	Bullet Average Velocity m/s (ft/s)						
	400 (1312)	500 (1640)	600 (1969)	700 (2297)	800 (2625)	900 (2953)	1000 (3281)
1 (3)	3 (11)	2 (7)	2 (7)	1 (4)	1 (4)	1 (4)	1 (4)
2 (7)	5 (18)	4 (14)	3 (11)	3 (11)	3 (11)	2 (7)	2 (7)
3 (10)	8 (28)	6 (21)	5 (18)	4 (14)	4 (14)	3 (11)	3 (11)
4 (13)	10 (35)	8 (28)	7 (25)	6 (21)	5 (18)	4 (14)	4 (14)
5 (16)	13 (46)	10 (35)	8 (28)	7 (25)	6 (21)	6 (21)	5 (18)
6 (20)	15 (53)	12 (42)	10 (35)	9 (32)	8 (28)	7 (25)	6 (21)
7 (23)	18 (63)	14 (49)	12 (42)	10 (35)	9 (32)	8 (28)	7 (25)
8 (26)	20 (70)	16 (56)	13 (46)	11 (39)	10 (35)	9 (32)	8 (28)
9 (30)	23 (81)	18 (63)	15 (53)	13 (46)	11 (39)	10 (35)	9 (32)
10 (33)	25 (88)	20 (70)	17 (60)	14 (49)	13 (46)	11 (39)	10 (35)

Image 4: This table takes the Bullet Average Velocity (X-axis) and the Target Lateral Velocity (Y-axis), and **outputs the windage hold or dial (cell contents) in mils and MOA.** Having different bullet velocities for one rifle is important because average velocity falls as a function of distance. Therefore, marksmen would want to keep an additional table that shows Bullet Average Velocity as a function of target distance. Note that without a fast bullet, a slow target, or a robot pulling the trigger, even relatively slow-moving targets are impractical to hit at long distances. The windage hold from this table is simply added to the windage hold for wind.

However, because distance to the target increases a bullet's flight time and decreases angular velocity, these two formulas can be simplified:

▸ *(T Angular V in Mils)* × *(Bullet Time)* =
(((T Linear V) × *1000)* ÷ *(D to T))* × *((D to T)* ÷ *(Bullet V))* =
((T Linear V) × *1000)* ÷ *(Bullet V))* = *(Windage in Mils)*
(Windage in Mils) = 1000 × ((T Lateral Linear V) ÷ (Bullet V))
(Windage in MOA) = 3438 × ((T Lateral Linear V) ÷ (Bullet V))

In other words, distance to target *(D to T)* cancels out. Put another way, if a bullet takes twice as long to arrive, the target has twice the amount of time to arrive at the point-of-aim and the lead is twice as large. The result is:

▸ **(Windage in Mils)** = **1000** × **((T Lateral Linear V)** ÷ **(Bullet V))**

The turn value can be included as a separate variable as well:

▸ *(Windage in Mils)* = 1000 × *((T Actual Linear V)* × *(Value))* ÷ *(Bullet V))*

▸ *(Windage in Mils)* = (1000 mils × 1 m/s × 0.7) ÷ 1000 m/s = **0.7 mils**

Note that **the units for target velocity and bullet velocity do not matter** because they cancel each other out, but the units do have to be the same (e.g., both are m/s, or both are mi/h, but one cannot be m/s while other is mi/h).

To prepare to shoot at moving targets, a marksman could create tables with precomputed values, such as: X-axis = *(T Lateral Linear V)*, Y-axis = *(Bullet V)*, cell contents = *(Windage in Mils)*. Similar tables are used to account for wind speed very often. (See Image 4, Pg. 15.)

Leads can quickly become impractical as targets get farther because the margin of error shrinks with distance. For example, for a man with a chest depth of ⅓ m (1 ft) who walks at 1 m/s (3 ft/s), their chest only exists in a particular spot for ⅓ of a second. Shooting with this level of time accuracy is relatively easy when the target is only 10 m (30 ft) away and 1 m (3 ft) represents a large angle. However, as the target gets farther from the marksman, the angular width of a target decreases, and the opportunity for something to change during the flight time increases. Therefore, the range for shooting at moving targets is typically limited from 50 m or yd where a lead is useful, to 400 m or yd where a lead is practical.

Once the lead is determined it is added to the elevation and windage holds or dials to determine the final point-of-aim. Then, a marksman can either "track" or "ambush" the target. "Tracking" is to follow the target with the point-of-aim, moving the rifle as the target moves. "Ambushing" is to place the point-of-aim in front of the target and wait for the target to move into position. (See Image 7, Pg. 17.)

To **ambush** a target, a marksman aims at a point in front of the target's direction of movement. Then, the marksman fires when the target is away from the point-of-aim by the lead distance. Thereby, the target moves into the path of the bullet.

Ambushing is the easier technique because the marksman can stay still while shooting. However, marksmen who use the ambush technique tend to jerk their triggers because their timings must be very precise. Further, winds may have changed while a marksman was waiting, changing the windage.

Shooting a Moving Target

Image 5: A running deer is one of the most common moving targets. White-tailed have been recorded sprinting at speeds of 60 km/h (40 mi/h) and sustaining speeds of 50 km/h (30 mi/h) over distances of 5-6 km (3-4 mi).

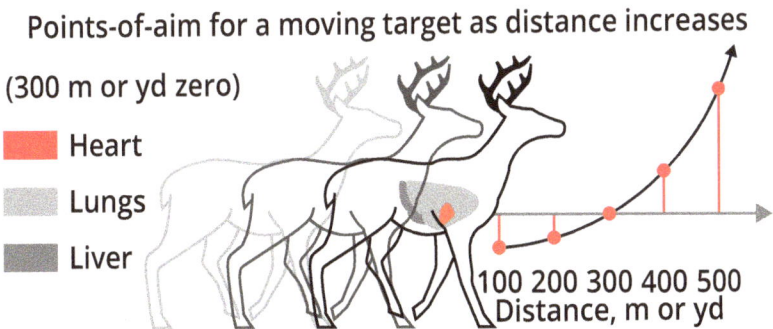

Points-of-aim for a moving target as distance increases

(300 m or yd zero)

- Heart
- Lungs
- Liver

100 200 300 400 500
Distance, m or yd

Image 6: As targets get farther from the marksman, a bullet drops more and takes longer to reach a target. Because bullets drop exponentially more with distance, the marksman must aim higher and higher. However, because the angle is constant, the horizontal offset increases linearly. Therefore, as targets get farther, the marksman's point-of-aim follows this upward curve as a function of distance.

Ambush, Trapping   ## Tracking

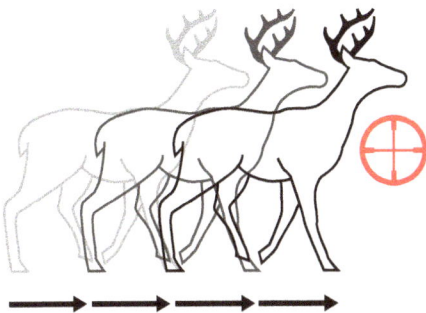

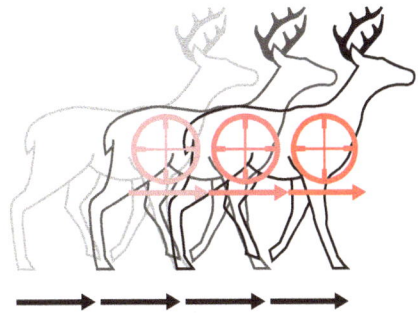

Image 7: A marksman can either hold their aim and wait for the target to move into position (i.e., **ambushing**) , or they can move with the target (i.e., **tracking**).

The **tracking** method moves the point-of-aim with the target so that it is one lead distance ahead of the target at all times. This technique allows the marksman more opportunities to fire, and they can be more patient. However, it is a more difficult technique because it requires moving the rifle while precisely aiming. While tracking, the marksman does not stop to shoot; the marksman pulls the trigger while continuing to move the rifle with the target and re-engages the sight picture to where they expect the target to be.

Tracking with a reticle in a perfectly straight line while maintaining a lead, windage, and elevation hold is not easy. In fact, focusing too much on the target and not the reticle is a major issue with marksmen who use the tracking method. Another danger of the tracking method is that what is behind the target changes, and so the marksman must pre-clear the entire potential background behind the target to ensure they don't shoot something unintentionally. A third issue is that if the rifle is moved too much, the marksman may move themself into a bad shooting posture.

To address some of these issues, a marksman can improve their tracking by practicing moving the reticle in a straight line. And they can position their rifle on a single pivot point such as a sandbag (instead, for example, a tripod or bipod). Further, the marksman must practice pivoting their entire body if the necessary swing is more than a few degrees.

A great method is to combine the two techniques. A marksman can begin with an ambush because it is better for maintaining a stable position. But the marksman is then prepared to immediately track the target if they miss their window of opportunity to ambush.

1.b Moving Marksman

The concept of a lead was explained in the previous section as a way to hit a moving target. (See Moving Target, Pg. 10.) That lead is a positive lead since it requires aiming in front of the target's direction-of-movement.

In contrast, when the marksman is moving, they must apply a negative lead, which requires **aiming behind the target**. A negative lead is required because when a marksman is moving, their rifle and ammunition move with them. Therefore, a fired bullet retains the marksman's sideways momentum, causing it to move sideways from the initial point-of-aim.

The concepts of positive and negative leads make intuitive sense when combined. That is, if both the target and marksman were to move in sync, then the negative and positive leads would cancel each other out and the marksman could fire without using any lead at all.

Shooting from a Truck

Image 8: In this scenario, the bench has been lowered to create a shooting support. Island of Crete, Greece, 12 Mar 2018.

Image 9: While a **gimbal stabilizer** could be used, if a marksman can plan for a gimbal they can usually plan a way to shoot without being in motion.

Image 10: Italian special operations forces engage targets from a moving truck. They spread their legs to **find stabilizing footholds**. Crete, Greece, 18 Mar 2018.

Calculating a negative lead is the same as calculating a positive lead, except the resulting hold is in the opposite direction of movement. In fact, it is quite a bit easier since often a marksman precisely knows their own movement speed, unlike having to guess the speed of a moving target. Therefore, it is even possible to shoot out the back of a fast-moving vehicle. (Again however, the easiest way to shoot from a moving position is to simply stop moving.) For example, the final formula for a lead was this:

► *(Windage in Mils)* = 1000 × ((*T Actual Linear V*) × (*Value*)) ÷ (*Bullet V*))

(*T Linear V*) can be replaced with -(*M Linear V*) for negative marksman linear velocity. That gives the formula:

► *(Windage in Mils)* = 1000 × -(*M Actual Linear V*) × (*Value*) ÷ (*Bullet V*)

For example, assuming that a vehicle is moving laterally at 30 k/h (18 mi/h), or -8.3 m/s (-9 yd/s) after converting so that the vehicle velocity and the bullet velocity use the same units, and assuming that the bullet is traveling at 1,000 m/s (1093 yd/s), then the formula looks like this:

- *(Windage in Mils)* = $1{,}000 \text{ mils} \times (-8.3 \text{ m/s} \times 1.0) \div 1000 \text{ m/s} = \textbf{-8.3 mils}$

Because the bullet moves sideways at the speed of the vehicle, **even moderate vehicular speeds can quickly require a large amount of windage.**

The most difficult aspect of shooting while moving is **muzzle vibration**. There are a variety of methods that have been devised to stabilize a rifle on a moving platform: bag supports, single straps, and spider straps are a few examples. A gimbal may be used, but the scenario where a rifle gimbal would both be useful enough to plan for, but the shooting task couldn't be reconfigured to not involve vehicular movement is unclear (See Image 9, Pg. 19.) Whatever the support used, a rifle cannot be stabilized directly against a vehicle frame because vibrations easily transfer from metal to metal.

The **bag support method** is the fastest to establish and provides a stable shooting platform. A soldier can utilize a variety of bags, such as: a rucksack, an assault pack, or a luggage bag. The more firm cushioning, the better. It is crucial to ensure that the bag itself is secured firmly to the interior floorboard of the platform to guarantee stability and accuracy.

The **single strap method** is also reasonably quick to establish. First, a marksman needs a readily available, inelastic line such as 550 cord, a rifle sling, or even a roll of duct tape if desperate. Then, the marksman secures either end of the line to the left and right sides of the vehicle frame, creating a horizontal shooting platform. The line must have some degree of slack so that the rifle can hang in the center of the line. To stabilize the rifle, a marksman leans the rifle into or away from the line, so ideally the line is strong enough to support many times the force of an adult man. (See Image 11, Pg. 21.) (See Image 12, Pg. 21.)

The **spider strap method** is the most complex method, but also the most stable. This technique uses four attachment points instead of the two used with the single strap method. To make the spider strap, a marksman attached two or more lines to the top left, top right, bottom left, and bottom right corners of the vehicle's opening (e.g., a doorway). The lines converge at the center, where they are held together by a carabiner, tape, or another locking device. A cushion can also be attached to the center to help dampen vibration. (See Image 13, Pg. 21.) (See Image 14, Pg. 21.)

To reiterate, the best way to improve accuracy while shooting and moving is to reduce vibration. To do this, the marksman can position themselves on a

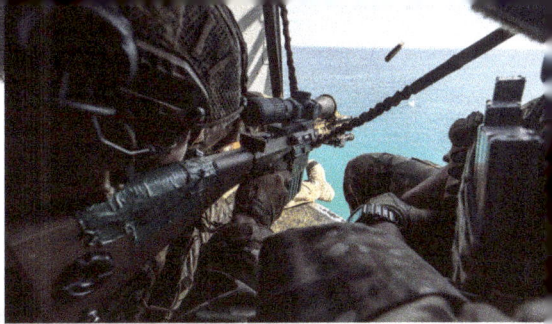

Shooting from a Helicopter

Image 11: A Soldier from Alaska trains with a M107 .50-Caliber Long Range Sniper Rifle in a UH-60 Black Hawk. Note the string is wrapped around the rifle. Joint Base Elmendorf-Richardson, AK, 07 Mar 2014.

Image 12: A U.S. Army Special Forces Sniper engages a floating target from a Greek Navy Bell UH-1 Iroquois (Huey) helicopter. He is pulling his line slightly back for increased stability. Souda Bay, Greece, 21 Apr 2021.

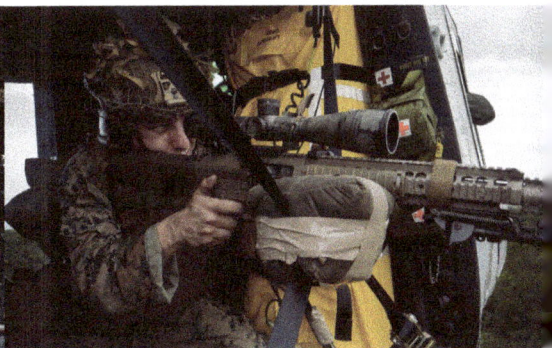

Image 13: A U.S. Marine corporal practices maneuvering a M110 semi-automatic sniper system inside a UH-1Y Venom helicopter. Camp Lejeune, NC, 12 Mar 2015.

Image 14: A U.S. Marine Corps lance corporal scout sniper engages a target with an M110 semi-automatic sniper system while aboard a UH-1Y Huey. Okinawa, Japan, 14 Jun 2022.

cushion. They can also use the scope on the lowest magnification setting so the vibration does not appear as significant in the sight picture.

Shooting from a moving vehicle at any significant distance (e.g., beyond 100 m or yd) has a high chance of requiring multiple shots to hit the target. To that end, a semi-automatic rifle is much faster to reload than a bolt-action.

Finally, **shooting from a moving platform is a team effort** between the marksman, driver, and possibly a spotter. If the marksman needs a different speed or direction, they can always ask the driver to change course or hold still. In fact, a vehicle can hold multiple snipers, a sniper team leader, spotters, and a driver to create a long-range shooting vehicle. Having multiple marksmen aiming at the same target is a sure way to increase the odds of at least one successful hit.

2. Multiple Participants

A typical shooting interaction has one marksman and one target. However, there are many instances in which there are more people or targets involved in a single scenario. For example, if a marksman has too much work, they can offset some of that work onto an assistant called a "spotter." In other scenarios, there may be multiple targets, multiple marksmen, or even multiple targets and multiple marksmen. These complex scenarios may involve tens of people who all must **coordinate their actions with high precision**.

2.a Working with a Spotter

A two-person long-range shooting team is composed of a marksman and a spotter. The marksman is the man behind the gun, and their job is simple; the **marksman only needs to aim at the target and pull the trigger**.

In contrast, the spotter's job is more complex. They serve as the vital **information supplier for the marksman**. The spotter gathers information such as the wind speed and direction, and also the impact location of prior shots. With this information, they use the various charts, tables, and the ballistic calculator they are holding to generate elevation and windage holds for the marksman. Finally, for occupational marksmen with command teams, **the spotter is the point-of-contact** and handles all external communications.

Teamwork is highly effective because dividing the tasks among multiple people is faster than having one person handle all the work. This increased speed not only allows each shot to be taken more quickly but also improves accuracy. Increased accuracy occurs because dynamic factors in shooting, such as wind or moving targets, can be addressed before they change a lot.

Notably, given two experienced marksmen, the more skilled marksman is the spotter because they can give more accurate information faster. Aiming and pulling the trigger is considered to be the more straightforward function. That is, while being a marksman does require skill, the job requires fewer total skills than being an effective spotter requires.

Having a spotter enables the team to use additional optical equipment. Because spotters normally do not shoot, they can utilize binoculars or spotting scopes instead of riflescopes. Both devices can be held closer to the eye than riflescopes because riflescopes must be mounted far from the marksman's eye to prevent recoil impact. Thereby, compared to riflescopes, **binoculars see a wider area at the same resolution and spotting scopes see the same area at a higher resolution**. (See Image 15, Pg. 23.)

Working with a Spotter

Image 15: Two Marines from the 3rd Marine Regiment form a standard **sniper-spotter team** with a spotting scope and an M110 sniper rifle. The most common position for a spotter is right next to the marksman in order to facilitate easy communication. Fort Bragg, NC, 22 Mar 2021.

Image 16: A joint US and UK sniper team engages targets from a truck to elevate the sniper's vantage point while protecting the spotter. So long as **communication is reliable**, the spotter and marksman can be in entirely different positions. Camp Asazi, Latvia, 16 Nov 2021.

To begin shooting, the spotter and the marksman must agree on their communication and verbage. Communicating information in the fastest, least ambiguous way is vital to successful teamwork. The most important parts of communication start at target identification and end with the marksman shooting the target.

During shooting, **the first step in any situation is to find and identify the target**. Once the spotter determines what the next target is, they must be able to direct the marksman onto that target. Many methods are available; for example, a common method in a well-planned area is the 3D's: a direction, a distance, and a descriptor. For example, "278 degrees at 500 m, male with red shirt and jeans." Another, rougher method is for the spotter to use clock degrees relative to the direction the marksman is facing: "Your 1 o'clock at approximately 500 m, male with red shirt and jeans." A spotter can also direct a marksman from an obvious landmark in the area: "From the intersection look 50 m to the East, male with red shirt and jeans." Any target description must be so specific and distinctive that it only applies to a single possible target! The marksman responds once they find the target: "male with red shirt and blue jeans, ready."

After confirming the target, the spotter reminds the marksman to complete a **basic checklist** to eliminate any chance that a vital step could be forgotten. (The demographic most likely to accidentally skip vital steps are experienced teams that become complacent!) An example is, "laze, parallax, mil":

Laze – Use a laser rangefinder to determine the exact distance to the target.

Parallax – Adjust the scope's parallax setting to match that distance.

Mil – Measure the height of the target in mils or MOA.

Once the target and its characteristics are identified, the spotter **calculates the elevation and windage holds** using their tools, such as charts and calculators. (Premade tables are manufactured using ballistic calculators prior to the shooting event.) If technology is unavailable, the spotter can ask the marksman for their estimate and average it with their own, as averaged estimates are usually more accurate than any individual's estimate.

Then, the spotter gives winds and elevation holds to the marksman. A hold refers to distance on the marksman's reticle, so for example, a spotter may say, "Left 2, Up 6.8" to tell the marksman to move the center-of-the reticle left 2 mils and up 6.8 mils (i.e., the target is centered in the bottom right quadrant 2 mils to the right and 6.8 mils down from the reticle's center). **The marksman always confirms**, "Left 2, Up 6.8, Ready." Regardless of the number of inputs, the call to the marksman is always the same: a single left-or-right number and a single up number. A spotter does not give feedback in clicks, nor do they give guesstimates for the marksman to interpret. That said, a spotter must not hesitate to update their instructions if something changes that requires a new call, such as a gust. **Spotting is a continuous process.**

The spotter and marksman speak the fewest number of words possible to maximize verbal clarity. Some spotters even prefer to only give one direction and number at a time. However, it can be a good idea to add the word "point" before the direction, for example, "point left 2." This additional word ensures that the marksman understands that the given direction is where the marksman must point the center of the crosshairs (i.e., to the left), and that the marksman does not instead mistakenly put the target in the reticle's left quadrant (i.e., pointing the crosshairs to the right). While some hunters work with spotters that give information in less compressed form, such as giving the wind speed itself instead of a wind call, this is only to entertain the marksman and is not a maximally effective setup.

Spotters base their corrections on the impacts of previous shots. Therefore, a marksman must either follow the spotter's instructions precisely or refrain from acting altogether (for example, declining for safety or pausing for verification). In other words, **the marksman must not modify the spotter's directives**. Any deviation by the marksman causes the spotter to adjust based on their original, ignored instructions rather than the marksman's actual shooting, leading to baseless adjustments. For instance, if the spotter calls "left 2.2 mils" but the marksman only adjusts left 1.1 mils, subsequent instructions from the spotter would be based on the 2.2-mils instruction instead of the reality.

Image 17: A Marine sniper attached to the 24th Marine Expeditionary Unit **receives information from his spotter** behind him as he shoots at a target in the ocean during a crew-served weapons shoot. USS Nassau (LHA 4), 12 Feb 2010.

Sample spotter-sniper interaction:

Spotter – identifies the target.

"From building 4, go 9 o'clock 10 mils. One male, red shirt, blue jeans."

Marksman – confirms the target.

"One male, red shirt, blue jeans, standing next to a white Hilux truck."

Spotter – runs through the checklist.

"Check parallax, laze, and mil."

Marksman – confirms ready; gives the requested information.

"Ready. Height, 2.0 mils."

Spotter – gives the elevation hold.

"Point up 6.2 mils."

Marksman – confirms the hold.

"Pointing up 6.2 mils. Ready."

Spotter – gives the windage hold.

"Point left 0.5 mils."

Marksman – shoots.

Shoots.

Once a bullet impacts, the spotter incorporates the information into a new call. They do not inform the marksman of anything other than "hit" or "miss."

Reporting unnecessary information back to the marksman can only distract them. Even worse, telling the marksman where they missed allows for conflicting opinions in the team on what a good adjustment would be.

A miss can occur due to either incorrect instructions from the spotter or poor performance from the marksman. If the marksman knows that they shot poorly, they must inform the spotter so that the spotter knows their instructions do not need to be changed. Then the marksman can try again using the same windage and elevation holds.

A marksman explicitly indicates the direction of their mistake if they are confident. For example, "bad shot, 6 o'clock." A marksman may also indicate that they shot well with a phrase such as "good."

If neither the spotter nor the marksman cannot identify where an impact occurred, the spotter can give the marksman the same holds to repeat the shot. (When bullets leave no signs of impact, they typically are going high.) A spotter may also use the same hold again to verify that the marksman is shooting consistently. If the same holds lead to different impacts, the marksman and their rifle must both be inspected.

If a spotter is working with different marksmen at different times, they must consider that every marksman responds to information differently. How elevation and windage holds affect shots depends on the marksman's rifle, scope, batch of bullets, and interpretation of commands. Therefore, even if two marksmen appear identical, the same hold command can lead to different results.

Spotters are particularly relevant to the military and police because those professions place additional administrative duties on a marksman that are external to actual shooting. Such duties include: communicating with higher command, maintaining situational awareness, terrain navigation, and recording events. If long-term observation is required, the marksman and spotter can switch positions to reduce their eye strain. Military teams can also carry diverse loadouts, with a team carrying a bolt action rifle for sniping and a semiautomatic rifle for the spotter as a medium-range backup weapon.

All that being said, modern technology is obsoleting spotters because modern technology makes it easier for a marksman to gather their own information. For example, shooting competitions use cameras at the target to determine impact. Ballistic calculators, better reticles, and more aerodynamic bullets increase the chances of first-shot impacts at distances that would have been extremely unlikely in past years. And communication is easier with headsets, cellphones, and digital readouts. For example, in the past a command team had to relay orders to multiple units by handheld radio, and those units then had to verbally report information back up the chain-of-command. Now, marksmen can receive commands through an earpiece, and transmit their points-of-aim and sight picture in real time by using scopes with a broadcasting video feed.

This influx of technology has not completely depreciated the role of a spotter in the military and police, but rather, the new technology has changed the spotter's role. Sniper-spotter teams are still teams of two; however, neither marksman is a dedicated spotter and both can either shoot from their own rifles independently, or they can work together to shoot one rifle more accurately. This flexibility doubles the potential firepower of the team. (See Image 18, Pg. 28.)

Technology also allows the spotter to be in a **separate location** from the marksman. (See Image 16, Pg. 23.) Being in separate positions can be useful because it allows the marksman to shoot over an obstacle to hit a target that they cannot see. To perform this shot, the spotter picks a proxy target that is close to the actual target, such as a bright rock, and describes it to the marksman. The marksman-spotter team shoots the proxy target until they can hit it consistently. The spotter then proceeds to determine the angular distance between the proxy target and the actual target. The spotter tells the marksman to always aim at the proxy target and dial that angular distance into their scope. As the marksman always aims at the same spot (i.e., the proxy target) changing the dial by the angular distance to the proxy target makes the bullet impact on the proxy target. If the first shot does not impact correctly, the spotter can guide the marksman to impact the actual target by having them dial their scope while continuing to aim at the same location. This process is similar to how a forward observer would guide artillery.

All that said, being in separate locations is usually a bad idea, as it increases the vulnerability of the marksman. A primary role of a spotter is **situational awareness** to ensure the safety of the marksman while they shoot.

Although the most common military teams only employ one marksman and one spotter per team to conserve manpower, competition shooting may not be constrained to a two-man team. One marksman can assign multiple spotters to observe and report on different environmental factors individually, such as one on bullet impact and one on wind changes. Similarly, a single spotter can set up two spotting scopes to switch between, one to observe the target and one to observe mirage. In fact, a competition marksman can use another competitor as a quasi "spotter." The first marksman can wait for their competition (of equal or greater skill) to shoot first, and then use their impacts to better understand the wind conditions.

Multiple Marksmen

Image 18: A U.S. Army Special Forces sniper team fires upon a floating target at sea. While spotting scopes can see targets in more detail, often **another scope can perform the same function**. Using two rifles allows the spotter to be a backup marksman. Souda Bay, Greece, 20 Apr 2021.

Image 19: **One spotter facilitates two competitors** in the USASOC International Sniper Competition. And vice versa, multiple spotters can facilitate one marksman. For example, one determining wind speed and another looking for impacts. Fort Bragg, NC, 22 Mar 2021.

2.b Multiple Marksmen

A marksman only has one chance to hit a living target before that target moves to cover, be it an animal or an enemy. However, this limitation can be overcome by using multiple marksmen who all shoot at the target to **increase the odds** that at least one bullet impacts correctly. Marksmen do not have to be collocated if they can still communicate via radios.

Every marksman involved in the simultaneous engagement must initially verify that their target is in sight and that they are ready to execute the shot. This confirmation is typically delivered through verbal notification. However, modern camera technology may allow the marksman to wirelessly send their scope's sight picture to other devices. This technology allows a spotter or team leader to directly see whether the target is in sight without asking.

To **facilitate synchronization of shooting**, a team leader or spotter often counts down, such as by saying, "Standby. Ready. Fire." Or a team may use a reverse order countdown, "Three. Two. One." Although shooting on the "one" is possible, even that duration of sound is often considered too imprecise. Therefore, snipers can be instructed to shoot at the precise sounding of the "T" in "Two." A third option is to use an electronic audio recording with three beeps. Ideally, all the shots from every marksman are fired so closely in time that they sound like a single shot. Marksmen can practice their coordination with dry-firing, so that the clicks of their firing pins sound like one click.

Whatever trigger is used, marksmen must never predict the countdown. If the trigger word is not said at the end (e.g., "fire" or "two") it is likely because there was an interruption and the team leader does not want the shots to be fired. In fact, some teams utilize **abort-codewords** specifically to avoid inadvertent shooting.

Another common pitfall to avoid is jerking the trigger. Because marksmen are so focused on precise timing, they may jerk their trigger at the last moment. It is important to remember that both no shot and a delayed shot on target are preferable to any missed shot.

2.c Multiple Targets

Sometimes, a single marksman has to shoot at multiple targets. For example, a military sniper may be responsible for suppressing one side of a building with many windows. A tournament competitor may have to hit multiple targets in quick succession. Or a hunter may spot multiple animals that they want to engage to bring home as much meat as possible.

The first step in planning to shoot multiple targets is to **prioritize and order the targets**. This process always begins with defining the area of engagement. For example, the marksman may be responsible for a building or a ground sector. Hunters may want to monitor one field more closely, or multiple fields more broadly. For complicated terrain, the marksman can create a map or chart with specific distances listed. In contrast for simpler sessions, the marksman may simply use estimates and a rangefinder in the moment.

Within their decided area, the marksman or marksman-spotter team **systematically reviews** the area for potential targets. For example, when scanning the side of a building, every opening (i.e., window, door, or hole) is examined from the top left to the top right, and then each floor below in succession. A common method of labeling for buildings is to label the top left opening "1A," with letters going to the left and numbers going down. The top floor is floor "1" because the bottom floors are not always visible. (See Image 20, Pg. 31.) When a potential target is identified, it is noted.

The second step in shooting between multiple targets is **creating an order of priority**. There are multiple methods a marksman can use. Target marksmen often diagram their targets and shoot them in the order of closest to farthest. Using a diagram enables the marksman-spotter team to refine their communication and get in the groove of shooting as they move along to more difficult shooting.

Military snipers often engage targets by eliminating the most serious threats first. "Threat" can be defined by either danger or reactivity. That means that a sniper would first engage the targets that are the most dangerous and the most likely to react to prior shots. This prioritization both minimizes the danger and ensures that as few targets flee as possible. If an extreme threat is identified mid-shooting, a military marksman may even pause to reassess their shooting order to take immediate action against that extreme threat.

Hunters with multiple targets have the goal of taking home as much prey as possible. To that end, "a bird in the hand is worth two in the bush." In other words, hunters take down the easiest targets first. That said, if the herd's lead doe is both available and obvious, shooting them first can sometimes create confusion in the herd, potentially creating a brief window to shoot again. Similarly, shooting a deer on the fringe of the herd may also delay a complete scattering. However in all other cases, attempting to shoot some animals to influence the fleeing direction of the remaining animals can be an exercise in futility. Finally, trophy hunters always prioritize the trophy animal.

Once the marksman has determined where their targets are and in which order they would shoot them, the **marksman must get into position**. The chosen shooting position must be able to accommodate transition between targets. For example, the prone position limits lateral mobility and is not ideal for shooting short-range targets over a broad area. In contrast, a kneeling position is not stable enough to engage multiple targets far away. The standing position with a support tripod is a good choice for engaging multiple targets at a long range because it provides support and stability while also providing room to observe and transition. However, instead of locking the rifle into the tripod, the marksman can rest the rifle on a sandbag situated on top of the tripod to enable easy adjustments.

When shooting and transitioning between targets, a marksman must be **methodical**. Some military marksmen fire two rounds at every enemy to ensure that at least one hits the target. Other military marksmen fire one round at every enemy as quickly as possible to put down as many threats as quickly as possible. Afterwards, they return to each threat to reassess and shoot again. Hunters in contrast, may fire only one round and re-engage their sight picture before firing a second shot in order to preserve as much meat as possible. Target marksmen engage in whatever method their competition prescribes.

Reducing the time between transitions is difficult and takes practice. One method of reducing time is to maintain a tempo. Keeping a tempo between shots encourages the entire movement to become methodical (e.g., moving the left leg, then the right leg, then the hips, etc.) Keeping tempo also encourages a marksman to not rush their shots, which can cause the marksman to jerk their trigger. Of course, using a semi-automatic rifle and practicing shooting are also both imperative for becoming faster.

Window and Door Coordinates

Image 20: When marksmen and spotters communicate about buildings, they must be able to **quickly identify windows and doors**. This is done by starting at the top left corner because the bottom of the building may not be visible. If multiple wings, faces, or frames are visible, they too must always be specified.

Image 21: In this example, the highlighted area could be fully identified with: "East wing, East face, window 1C, bottom frame." (Doors and windows are counted separately.) Alternatively, to **prevent confusion** about wings on this four-story building, the area can also be identified with: "East face, window 3C, bottom frame."

2.d Multiple Marksmen and Targets

Rarely, multiple targets must be taken out simultaneously, leaving no time for any target to react. This exact scenario occurred when Captain Phillips was held hostage in a lifeboat by three Somali pirates. Navy Seals had to eliminate every pirate before they could react and take revenge by killing Captain Phillips. In the movie reproduction of this event, the Seals took three simultaneous shots to confirm three kills. In real life, it is unclear whether the Seals were as perfect in their synchrony, but the scenario and successful rescue of Captain Phillips were the same nonetheless.

Coordinating multiple marksmen to shoot multiple targets simultaneously is not much different from coordinating multiple marksmen to shoot one target. Therefore, the individual instructions from a spotter to their marksman are the same. (See Multiple Marksmen, Pg. 28.) And assigning multiple targets to multiple marksmen is done the same way as assigning them to a single marksman. (See Multiple Targets, Pg. 29.)

The new aspect of coordinating multiple marksmen and targets is the introduction of a team leader. **Having a single, centralized team leader is vital to ensure proper coordination among various, dispersed teams.** Team leaders must ensure that each spotter is entirely sure of their assigned target using simple, precise language. This clarity can be created by using only a single reference point for all the shooting teams. For example, on a building, that reference point could be the top left corner of the building. Then, each shooting team is assigned different coordinates in relation to that single reference point.

Also, because each target may appear or disappear at any time, each team must send up reports on their readiness to fire on the target whenever their situation changes. Traditionally, these reports would be sent up in a **round-robin report.** That is, the team leader would ask for a status report from each spotter, and the spotters in order would report their identity and status, such as "Team 1, ready." or "Team 1, no go." In contrast, modern spotters can send a video feed of their sight picture to the team leader so the team leader can view the status of all marksmen simultaneously.

Once the team leader is satisfied, he can issue the preplanned trigger word, such as "engage." Or he can order the spotters to wait on whatever other trigger has been preplanned, such as a specific start time.

Positioning is also vital. The team leader must weigh the benefits of disbursing the shooting team to achieve better angles of fire against the costs of worse communication, security, and responsiveness.

Image 22: On February 4, 1976, near the Djibouti–Somalia border, a group of independence militants held a busload of children hostage. The GIGN, a French counter-terrorism unit, deployed to the scene and introduced a new tactic known as **"simultaneous shooting."** After careful coordination, each sniper targeted a different hostage-taker inside the bus, waiting on a precise, silent three-second countdown triggered by the command "zero." When they fired in unison, only a single shot was heard. All four captors fell instantly, and the children emerged unharmed.

Caption: Djibouti 1976: 1. Somalian MG 42. 2. Somalian border post. 3. Bus with the thirty child hostages. 4. Border barrier. 5. GIGN firing position. 6. French GNA fort. 7. Command post of General Brasart. 8. Legionnaires' starting point. 9. Armored car starting point (800 meters).

3. Restriction-Based Positioning

Marksmen want to be as accurate and precise as possible. So they tend to always use the same, optimal shooting positions. However, sometimes there is a physical restriction that prevents being optimal and the marksmen must adapt and make the best out of a suboptimal position. These are some of the more common scenarios.

3.a Non-Standard Shooting Positions

Marksmen are often trained to shoot from standard positions, such as prone, kneeling, and standing, as these provide the most stable and predictable platforms. However, **real-world scenarios are not in ideal conditions**, and marksmen must adapt to a variety of non-standard shooting positions.

Some of the more common non-standard shooting positions are sitting, squatting, and lying on one's back (supine position). Key features of non-standard positions include: unstable platforms, restricted space, and odd angles. Whatever the case, the end result is the same: there are fewer points-of-contact with the ground, less bone support (i.e., more muscular support) to the rifle, and less resistance to recoil.

Therefore, the first step in shooting from a restricted position is to **maximize the rifle's points-of-contact with the ground**. This means placing the rifle on a supporting surface whenever possible, even if the marksman themself must be in an awkward position. Of course, normal supports such as sandbags, bipods, and tripods are always useful. However if they are used, then the position is a standard supported position and not "non-standard."

Because locations that result in restricted positions are often characterized by unstable or uneven surfaces, such as rocky terrain or vehicle hoods, marksmen often cannot maintain a sight picture in these restricted positions for a long time. However, marksmen can extend the longevity of their sight picture by using objects in their surroundings as a support, such as local rocks and dirt, to elevate themselves or their rifles. Of course, physical training and endurance training can also both help a marksman to stay in an unexpected position for a longer amount of time.

Image 23: This marksman supports his rifle with a sandsock. **Rifle support** and stability are the foundations of every shooting position.

Image 24: This marksman is modifying a standing position. However he puts **no weight on the rifle** because forcing a rifle decreases its precision.

Image 25: Marksmen can shoot off **any stable surface**. It is important to not get tunnel vision, and incorporate unconventional scenarios in training.

Image 26: Sometimes when thinking of novel scenarios, it is helpful to look to **historical examples**. Many marksmen have shot from rooves.

Image 27: The **environment** may allow for positions that otherwise would be unstable. This slope makes sitting a much more stable position.

Image 28: Similar non-standard positions can be used in **both urban and rural** environments. This rock and a roof require similar positions.

Image 29: There are various ways of getting **as low as possible**, while still maintaining front-support for the rifle. This is a kneeling crouch.

Image 30: To get low, this marksman sits down and uses **the crux of his elbow** to support the rifle. He could also get low using the bipod.

Image 31: To get low, this marksman **lays on his back**. This is possible because his bulletproof vest and the tree are providing back support.

Image 32: Good **back support** can allow for a position that is almost as stable as a prone position. Trees and walls are common back supports.

Image 33: To match the height of the window, this marksman has to go lower than a crouch, but higher than a prone. So he **swings his legs**.

Image 34: If there is not enough space behind the marksman, they may have to completely **swing their body** to the side to get low.

Image 35: Often, **cover dictates** a marksman's position. This marksman, however, would be safer standing and using his left side to shoot.

Image 36: To compensate for his unavoidably poor shooting position, this marksman has **wrapped his sling** around the ladder to **lock in his rifle**.

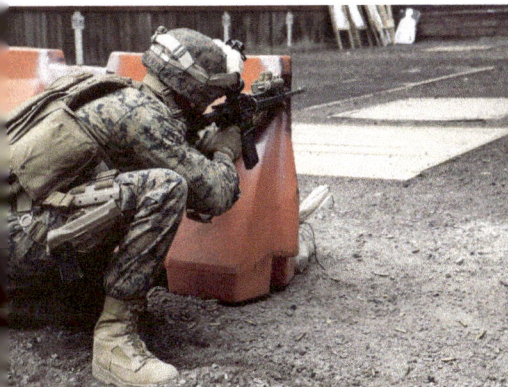

Image 37: Although this is training, the marksman has achieved **concealment but not cover** because plastic barriers are not bulletproof.

Image 38: Only the **engine block** is bulletproof, meaning the marksman on the left may be safer and more accurate in the prone.

Image 39: This is a fun trick shot. But there is **no reason** to have a marksman kneeling in a bare field, devoid of a real barrier or a support.

Image 40: Many times, marksmen practice positions because they are **fun and interesting** rather than optimal.

Non-Dominant Hand

Image 42: The most common scenario which requires using a non-dominant hand is when **cover is only available from one side**. In this diagram, the center figure shows what happens when a right-handed marksman fires from right-side cover. They must either find left-side cover (left figure), or switch hands (right figure).

Image 41: A U.S. Marine platoon commander with Special Purpose Marine Air-Ground Task-Force Crisis-Response Africa fires an M4 carbine around a high barrier using his **non-dominant hand** to maintain cover. He must be careful to not block the ejection port and have a hot casing get caught. Germany, 09 Apr 2016.

3.b Non-Dominant Hand

Ambidextrous shooting is a skill that many police and military snipers benefit from mastering. Training to fire a rifle from both sides of the body prepares marksmen to be able to **use cover on both their right and left sides**. In contrast, attempting to fire from cover that is on the side of the support hand leave's the marksman's torso exposed.

The most significant challenge when shooting from the less dominant shoulder is correctly positioning and maintaining the rifle against the shoulder throughout the firing process to mitigate excessive recoil. Also, maintaining correct eye relief with the non-dominant eye can be tricky. Both of these issues can be overcome with practice. **Practice is in fact the only answer**, as there is no trick or shortcut to becoming better at off-handed shooting.

It's advisable to start practicing non-dominant-side shooting from the prone position before gradually advancing to more challenging stances because it's typically easier to fire from the less dominant shoulder while in a prone position. Similarly, marksmen can **practice with dry-firing** before transitioning to live-fire practice to avoid wasting ammunition. Eventually, the goal is to be proficient at shooting from a standing, unsupported position from either side.

3.c Non-Vertical (Canted) Rifle

Rifles are held with the reticle perfectly vertically whenever possible. Introducing cant (i.e., sideways rotation) to a rifle causes bullets to travel in a complex three-dimensional trajectory that is difficult to predict or account for. However sometimes, a marksman has no other choice because they are shooting from a very restricted space, and both the scope and the barrel cannot point out of the opening at the same time.

If a rifle must be canted, a marksman is better served by rotating their rifle **either a full 90 degrees (i.e., the scope is beside the barrel) or only 45 degrees (i.e., the scope is diagonal to the barrel).** Doing this allows a marksman to more accurately adjust their elevation and windage holds. If the rifle is oriented at 90 degrees, the windage dial is adjusted with the elevation hold and vice versa, essentially switching the roles of the windage and elevation turrets. If the rifle is held at a 45-degree angle, the elevation dial setting is divided between the windage and elevation dials. Each dial receives about 70% (the sine and cosine of 45 degrees) of the elevation hold.

Every other angle requires more precise math to account for bullet-drop. Performing those calculations involves applying the sine of the cant times

Non-Vertical (Canted) Rifle

Image 43: A U.S. Marine Corps team leader with Special Purpose Marine Air-Ground Task-Force Crisis-Response Africa fires an M4 carbine through a nine hole barrier during a combat marksmanship range. Germany, 09 Apr 2016. Sometimes there is only a narrow slot from which to both aim and shoot through. In these cases, canting the rifle is necessary.

Image 44: Other times, **the position dictates canting**. This position may occur when shooting below a vehicle in a compressed space. On the other hand, it is unclear why such a rare scenario would be part of training. Germany, 09 Apr 2016.

Image 45: This Soldier is from the 11th Armored Cavalry Regiment. A more common, though still rare situation would be when a marksman must take a shot while pushing as far back as possible into **sloped cover**. Fort Irwin, CA, 04 Apr 2017.

the elevation to the elevation turret, and the cosine of the cant times the elevation to the windage. However, at that point, **it would be better for the marksman to just improve or change their shooting position to uncant their rifle**.

If a marksman somehow knows that they need to shoot through a very narrow opening, they can also address the cant of their rifle during the zeroing process. That is, a rifle can be zeroed with a cant. However, that zero point would then be invalid for the rifle in a vertical position, meaning the rifle could only be accurately be shot with a cant present, and only at a single distance (i.e., the zero distance). Instead, **a rifle can be "zeroed" with no zero point**. That means that the scope is adjusted to not account for the holdover, and is instead set so that the sightline is parallel to the boreline. This arrangement means that when aiming, the error introduced by cant cannot be greater than twice the holdover, which is a relatively small error when shooting at very long distances.

4. Disadvantaged Conditions

Basic shooting instructions assume that the conditions of shooting are optimal. The marksman is relaxed and not pressed for time. The sun is bright and warm. However, there are often circumstances that make a shooting session stressful, time-limited, dark, and cold.

4.a Stress and Fatigue

Physical stress can negatively impact both the accuracy and precision of shooting. An elevated heart rate, rapid breathing, and tremors or shaking in the hands make aiming more difficult. Then, high stress may lead to a rushed firing sequence, resulting in poor aiming practices. Finally, stress can cause a marksman to jerk their trigger more often, leading to inaccurate shots.

Fatigue, in addition to stress, can compromise critical cognitive functions such as high-level decisionmaking and the ability to perform calculations. For example, a tired marksman may choose to lay in the prone because it is more comfortable, despite another position providing a superior point-of-aim. Also, the precise calculations of factors such as wind, distance, and trajectory become more difficult.

Most importantly, **stress and fatigue can disrupt a marksman's established routine**. A marksman may skip specific steps in their routine because their memory recall is not as strong. Deviating from or rushing through any step in a standard procedure can significantly degrade performance in not just that step but also in every following step.

There have been a few scientific studies specifically done on the effects that physical stress and fatigue have on shooting, and they consistently show that the effects are significant but also minor and short-lived. For example, one study found that ingesting 300 mg of caffeine (a form of artificially induced stress) slightly decreased shooting performance from the standing position but not the prone. Another found that marksmen experienced a significant return to their baseline skill level after just three minutes of concentrated rest.

There are a few implications that can be learned from these findings. First, **taking a moment to deeply inhale and exhale** can eliminate most of the negative effects of stress and fatigue. That is because most of the errors can be traced back to rushing and not a decline in core performance. Another takeaway is that **forming habits and routines is important**, as stress and fatigue make marksmen rely more heavily on actions that can be performed without much active thinking or decisionmaking.

Practicing how to combat the effects of stress and fatigue is often misunderstood. The"skills" of calmness are to rely on previously learned skills and generally to slow down. That is, there is no special or different shooting technique that is optimal only for stressed and fatigued marksmen. Therefore, **making training stressful is a waste of time**, and a better use of time would be further ingraining core skills so that they appear when stress is induced.

That said, military and law enforcement snipers often do rigorous physical training and stress inoculation exercises. Introducing stress during shooting helps evaluate how marksmen might perform under pressure in real situations. (It's important to distinguish between training and testing, as they serve different purposes.) To test the stress reaction of these marksmen, they perform high-intensity physical activities followed by immediate shooting. Stress is not added to improve shooting accuracy, but to test how quickly the marksmen can calm themselves. Being calm allows a marksman to rely on their regular, practiced training routines even when stressed. Many marksmen achieve calm by focusing on and controlling their breathing.

4.b Time Pressure

There are many reasons why a marksman may be shooting under a strict time limit. For example, marksmen in competitions only receive a limited amount of time to complete their shots. For hunters or military, targets are often only briefly visible. For example, perhaps a deer may be eating grass, and the hunter can only see the deer's head when it raises it to look around. Or perhaps a military sniper only sees their target when they look out a window. There are also soft time "limits." For example, a marksman can improve their accuracy by only shooting during the portion of their time when environmental conditions, such as wind, are optimal.

The harsh reality is that sometimes a time limit simply does not give enough time, and a marksman misses their shot. However far more often, marksmen overcompensate and rush through their shooting when in fact, they still had remaining time to shoot in a slower, more controlled manner. The best marksmen learn to not miss their windows of opportunity by

Image 46: Two Soldiers, one Australian and one American, compete in a sniper competition. Rushing to get into position causes rushed and inaccurate shooting. It is more important to pause and remember that "**slow is smooth, and smooth is fast**" than to risk missing a shot by rushing. Camp Taji, Iraq, 29 Dec 2019.

following the **same deliberate process every time they shoot** to build up muscle memory. That is because quickness is the natural result of control and consistency. This idea is captured in sayings such as "slow is smooth, smooth is fast" and "consistency is key." In contrast, the aversion to forcing speed is captured in the sayings: "you can't miss fast enough" and "fast is fine, but accuracy is final." **The worst marksmen are those that shoot as fast as possible, without proper control.**

One of the best ways to improve muscle memory is to become more familiar with one's own equipment; for example, learning the specific way to manipulate the bolt of a bolt-action rifle. Marksmen who are familiar with their bolt can work the bolt quickly and quietly without looking at it. An even more advanced technique is for a marksman to time the moving of their bolt to their inhales and exhales. Another example of muscle memory is a deliberate trigger pull. Practiced marksmen know precisely how much force pulling the trigger requires and **are not surprised when the trigger releases**.

The shooting position also helps prepare a marksman for time pressure. For example, a marksman in the prone position cannot as efficiently change their natural point-of-aim (NPA) as a marksman in the standing position can change their NPA.

Image 47: A member of the Marine Corps Shooting team fires his rifle during the Quantified Performance Match. **The best way to prepare for time-sensitive shooting is to practice it.** Here, the marksman has his partner hold a beeping clock to enforce time standards. Quantico, VA, 24 Apr 2022.

Preparation is always necessary to shoot well under pressure. This rule applies not only to creating habits but also to the environment one is shooting in. For example, ammunition for follow-on shots must be placed within easy reach. Elevation holds for various possible target distances must be precalculated on a reference chart. The expected wind is logged and written down. A marksman can run through an entire scenario before it happens and perform every available action (save shooting) in advance of the actual event to save time.

Observing environmental factors can also help determine the true time limit for taking a shot. For example, if a target is about to stop moving, the marksman would know they have more time to aim carefully. If a marksman is relying on mirage to read the wind speed, they would benefit from noting how the clouds are moving and when they are likely to block the mirage. That is, environmental awareness can help such a marksman to anticipate periods when wind readings may become unavailable.

It's just as important to plan what not to do as it is to plan what to do. For example, if a marksman has a spotter to track previous shots, there's no need for the marksman to do it themself.

4.c Nighttime

Humans can't see well in the dark. It takes 30 minutes for the eyes and the brain to fully adjust to seeing at night. During this period, night vision gradually "comes alive," and everything seems to brighten. It's akin to the moonlight being turned up and the stars shining brighter.

The full moon can shine surprisingly brightly once one's eyes have acclimated to the dark. Under strong moonlight, a marksman with their naked eye can shoot pronounced targets within about 200 m or yd or less, with distances of less than 150 m or yd being ideal. Even a half-moon can illuminate some targets if they are large and prominent enough. In fact, hunters may align their schedules with the brightest phases of the moon if they plan on shooting at night.

That said, impacting targets over 200 m or yd away at night becomes very unlikely with eyesight alone. That is because **at night, eyes work differently** and lose much of their ability to see detail. This loss happens because low light does not activate some cells in the eye, disabling a portion of human sight. Specifically the eyes' cone cells, which handle central vision, color vision, and fine detail, do not work in darkness. That means that sight only occurs with the eyes' rod cells, which detect black-and-white vision and movement in the periphery (i.e., outside the center). As a result, although a marksman may sense a target's presence, aiming precisely at night becomes much more difficult. Eye fatigue also quickly occurs at night, after just a few minutes of intense watching. Therefore, some marksmen alternate eyes or plan for resting periods.

Another change in eyesight is **pupil dilation**. During direct daylight, pupils constrict down to 2 mm (⅓ in) in diameter, whereas at night they dilate up to 8 mm ($^1/_{12}$ in) (pupil size changes less with age). Pupil dilation dictates how large the most useful objective lens on a scope can be. A bigger objective lens creates a bigger exit pupil, (the exit pupil is the diameter of the light beam that passes through the back of a scope's eyepiece).

The size of a scope's exit pupil is determined by dividing the diameter of its objective lens by its magnification. For example, a scope with a 32-mm objective lens and 4x magnification has an exit pupil of 8 mm (32 mm ÷ 4). This 8 mm exit pupil matches the typical size of a human eye's pupil in low-light conditions, allowing the scope to fully illuminate the marksman's eye with light when the exit pupil is properly aligned. (To accommodate eye movement, good settings often have an exit pupil that is slightly larger than the size of the observer's eye's pupil.)

In contrast, a scope with 24x magnification and the same 32-mm objective lens produces a 1.33-mm exit pupil (32 mm ÷ 24). This much smaller exit pupil is insufficient to fully illuminate an 8-mm pupil in low light; therefore, the sight picture would be dimmer than the sight picture of a scope with a larger objective lens. This is why low-light scopes with high magnification settings can have truly giant objective lenses. This also means that it is not possible to judge the effectiveness of a nighttime scope in a store; real-world night conditions and fully adjusted night vision are required.

Some marksmen use flashlights, car headlights, or spotlights to add light to areas to night-shoot. Of course, adding light is effective at increasing the ability to see. However, if target shooting, operating in low-light conditions is adding unnecessary danger. And when hunting or on a mission, shining bright lights alerts the target to the marksman's presence and location. Therefore, adding light is usually not a good option.

Instead of relying on moonlight or manmade lighting, a marksman can use **night-vision devices that read the existing light and then amplify it** so they can see it. These devices can either be integrated into the scope itself or be a separate clip-on device that goes in front of or behind the scope. Most night vision devices are clip-ons. Because a clip-on night-vision device does not disturb a scope's mounting, the scope and rifle can maintain their zero both with and without the device. In contrast, marksmen with a scope that has integrated night-vision capabilities typically use a different rifle for daytime shooting with a daytime scope to avoid having to unmount and remount scopes. And even if mounting a clip-on device does shift the zero (a.k.a., a "zero shift"), the clip-on usually shifts the zero the same repeatable amount every time, so the marksman can simply dial or undial the zero-shift into the scope. The downsides of clip-ons compared to dedicated night scopes are the added weight, the added cost, and that both devices must be compatible.

Night vision devices generally use one of two methods: **image intensification (II) or thermal imaging (TI).** II devices see and amplify visible light. In contrast, TI devices see infrared light and display it as visible light. In other words, an II device displays how bright an object is; e.g., a cold white rock will show brighter than a warm black cat. (See Image 51, Pg. 47.) In contrast, a TI device displays how warm an object is because warmer objects emit more infrared light; e.g., a cold white rock will show darker than a warm black cat. (See Image 50, Pg. 47.) Some modern devices combine both II and TI into one image. Due to the niche and complicated nature of night vision devices, it is always necessary to read their manual. And all night vision devices require batteries, so carrying extra batteries is a must.

Image 48: A nighttime U.S. Air Force KC-135 Stratotanker. Al Udeid Air Base, Qatar, 09 May 2021.

Image 49: A daytime Nebraska KC-135 Stratotanker. Pardubice Air Base, Czech Republic, 14 Sep 2020.

Image 50: A thermal image of a KC-135 Stratotanker aircraft. MacDill Air Force Base, FA, 27 Feb 2019.

Image 51: An intensified image of a KC-135 Stratotanker aircraft. Al Udeid Air Base, Qatar, 02 Sep 2017.

Night vision devices display their output on a screen. Because screens have limited resolution, **when a night vision scope is magnified, no additional information is revealed**. However, more information may be perceived by the marksman since it is easier to focus on an enlarged area. And many night-vision devices are built with very high resolution specifically to allow for magnification.

An II device intakes light and converts and multiplies it into electrons that hit the back screen. That screen then displays an image to the marksman. Because II devices only intensify light, they rely on a difference in light between the target and the background. That is, II devices cannot illuminate; they can only intensify. Therefore, on plains under full moonlight, marksmen using even relatively old II devices can easily see and shoot large, reflective targets at 1,000 m or yd and beyond. However, if a marksman were shooting at a target inside a triple canopy forest under a new moon, the II device's effective range would be much shorter. That said, technology is continuously improving, and images are becoming crisper with each generation of II technology.

The advertised ranges of night vision devices can be deceiving. That is because night vision devices advertise the range at which a person can perceive the presence or absence of a target, and not the range at which a target can be accurately aimed at. For example, one high-end II device claims

Light Signature

Image 52: A U.S. Marine sergeant sets security. Night vision devices can **create a light signature** that can be detected in complete darkness. USS Essex, Pacific Ocean, 02 Sep 2021.

Image 53: A Marine with 3rd Battalion, 3rd Marines conducts a live-fire sniper range. Of course, the most obvious light signature is the firing of the rifle. Okinawa, Japan, 16 Aug 2022.

that it can "identify" a human at 1800 m but only recognize that human's characteristics at 450 m. In other words, with a target at 1800 m, the observer may only know whether the target is present or absent, which is not very useful for aiming nor shooting.

Although II devices are not primarily or explicitly thermal or infrared devices, most can interpret infrared light. Therefore, **marksmen often pair infrared illuminators (i.e., lights, spotlights, or floodlights) with their II device** to shine light onto an area that unequipped enemies and animals cannot see. Boar hunting with an II device and an infrared spotlight is relatively common. In contrast for the military, since II devices have become cheap, most militaries do not use infrared illuminators since they give away the marksman's location. Target marksmen also do not use illuminators since they could just shoot during the day if they wanted to add illumination.

Another potent combination for long-range shooting is to use an infrared laser as an aiming device. (See Image 54, Pg. 49.) The laser is attached to the rifle and runs parallel to the boreline. Thereby, when shooting at night, even if the target is difficult to see, the marksman can shine the laser onto the target to align their boreline and illuminate the target. Because the dot must reflect off of something to be seen, when the laser dot is seen on the target that means the boreline is pointing at the target.

If the target is relatively far away, the marksman can also apply windage and elevation dials to the laser to aim more precisely, just as they would for the scope during the daytime. The laser dial does not need to be seen because the dialing is confirmed through the scope. In contrast, the scope itself is rarely dialed at night for the simple reason that the dial markings cannot easily be seen in the dark.

Lasers Sights

Image 54: Paratroopers from the 82nd Airborne Division train using **infrared lasers and night vision**. Laser sights allow a marksman to illuminate their point-of-aim. Good laser sights can be dialed for elevation and windage hold, and have low light divergence so the laser dot stays small at long distances. Ramadi, Iraq, 26 Oct 2009.

Image 55: Unlike passive sights (e.g., a scope), laser sights are rarely mounted exactly vertically above the bore. In fact, **some are mounted to the side of the barrel**. Therefore, a marksman must know how far the actual point-of-aim is from the left or the right of the laser dot. East China Sea, China, 23 Sep 2020.

Another issue is that a laser increases diameter with distance, so at longer distances the dot becomes bigger and less precise as a point-of-aim. (This is called "laser divergence.") Therefore, it's best to zero a laser at 200 m or yd or less and then refine it at a second distance if necessary.

Although ideally a laser is mounted on the top of a rifle to horizontally center the laser with the barrel, there often is not enough mounting space to put one. Therefore, when zeroing a laser sight, **it is important to consider the laser's sideways-offset from the barrel**. For example, if the laser is placed 5 cm (2 in) to the right of the barrel, the marksman must always place the laser dot 5 cm (2 in) to the right of the actual point-of-aim. That is, the bullet and laser would travel in parallel and never intersect, similar to shooting with a canted rifle. (See Non-Vertical (Canted) Rifle, Pg. 39.) This offset prevents the laser from traveling sideways to meet the bullet, which would only occur perfectly at the zero distance.

An important factor to remember when using lasers in combat situations is their bidirectional nature. This means that while a soldier can see the laser, the target can also trace the laser back to the soldier, exposing the marksman's position. Another problem arises when multiple marksmen, all using laser sights, aim at the same target. Distinguishing one laser dot among many is difficult.

Thermal imaging (TI) devices only read infrared light. TI devices are heavier, bulkier, more expensive, and have less battery life than image intensification (II) devices. Although II devices also read infrared light, specialized TI devices are far more sensitive and accurate. Also, by not reading visible light, they can better reveal information that is only broadcasted through the infrared spectrum.

Just as the sun emits visible light (i.e., sunlight), **all objects naturally emit infrared light**. The amount of infrared radiation emitted is proportional to an object's temperature, and living organisms are typically warmer than their surroundings at night. Therefore, at night, humans and warm-blooded animals shine like beacons in the infrared spectrum. Even during the day, they continue to shine brightly in the infrared as long as their body temperatures exceed that of the environment, despite any high-intensity visible light. In contrast, if the surroundings become significantly hotter than normal body temperature, humans and animals still appear in the infrared spectrum, but relatively darker against the hotter background. When the environment is around body temperature, humans and animals become indistinguishable against the background.

Infrared light has different interactions with materials than visible light has. For example, materials that are transparent to visible light are often not transparent to infrared light. **Glass is opaque to infrared light**, and a TI device can only determine the glass's temperature and not the temperature or infrared image of whatever is behind it. In contrast, **dust clouds, smoke, and fog are all much more transparent to infrared light than visible light**. For another example, metal, wood, and paper targets are generally the same temperature as the environment and so can be difficult to distinguish in the infrared spectrum.

At night, finding targets in the first place is one of the most difficult steps in the shooting process. Therefore, a quality set of binoculars is crucial for target acquisition. Using binoculars with a minimum 40-mm objective after the eyes have adjusted to the darkness and moonlight can offer an astonishingly clear view without any artificial light. A very effective setup for hunting is to use a pair of II binoculars for detecting targets moving at night, then switching to a thermal scope to shoot.

4.d Cold

Cold can not only make for an intensely uncomfortable experience, but it can also directly change or degrade performance. Specifically, **the cold causes numerous performance issues** through changes in: material plasticity, thermal contraction, the water vapor capacity of air, battery performance, and a marksman's body temperature. Therefore, a marksman must consider the temperature of their shooting location during preparation.

Cold-Weather Clothing

Image 56: A U.S. Marine lance corporal tries to warm up his hands. Being cold is not just a matter of pain. Cold can cause uncontrollable shaking that ruins aiming. A marksman in serious pain or shaking from cold is a **failure of leadership** to plan and supervise. Sekiyama, Japan, 14 Mar 2017.

Image 57: Shooting in the cold requires a **full suit** of layered, waterproof clothing, boots, gloves, and headgear. On the other hand, excessive sweating is also bad. Sunglasses prevent snow blindness, and hiking sticks can double as shooting sticks. MT, 09 Feb 2023.

Changes to plasticity are straightforward: materials become more rigid and less viscous. Rubber materials such as gaskets are more brittle at lower temperatures and so are more likely to tear. Standard lubricants may thicken or freeze, hindering a rifle's performance. Switching to a lightweight, cold-weather-specific lubricant ensures smooth and reliable operation of moving parts.

Thermal contraction (or thermal expansion) is when a material changes physical size as the temperature changes. **Thermal contraction becomes a problem when there are multiple materials used closely together** in high precision. When it comes to scopes for example, metal housings dissipate heat more quickly than the glass lenses they enclose and therefore contract at a faster rate. This discrepancy in contraction rates can put tremendous pressure on the glass, leading to cracking in either the lens or the housing if the temperature shifts too rapidly. Thermal contraction may also pose a problem if a rifle is zeroed in a high-temperature environment but then fired in colder conditions because the rifle's overall shape changes slightly, leading to a corresponding small shift in the point-of-aim. Wooden stocks are the most prone to having this issue.

The third issue is that as air cools, it retains less water vapor. That means that if air inside of a scope is humid and cannot escape fast enough, cooling it down leads to water condensing inside of the scope. (See Image 58, Pg. 52.)

Image 58: This photo is of the shopping center Lauttis in Lauttasaari, Helsinki, Finland. It was photographed through a **fogged-up camera lens** because the rapid transition from exiting cold outdoor weather to entering warm indoor weather caused water vapor in the warm air to condense onto the outside of the camera. A similar effect can occur in reverse, **when a warm camera enters a cold environment and the warm air in the camera suddenly cools and condenses to water on the inside.** While exterior condensation can just be wiped off, internal condensation is much harder to fix, and often just requires time to acclimatize.

The problems caused by thermal contraction and water vapor capacity can mostly be resolved by allowing scopes to **slowly acclimate** to the ambient temperature. Transporting a rifle in the bed of a truck or otherwise exposing it to the cold over an extended period before shooting are ways to do this.

Electronic devices such as rangefinders and scopes experience **reduced battery efficiency** in cold weather. This reduced efficiency is caused by cold temperatures slowing down chemical reactions. To mitigate this problem, spare batteries are best kept inside a coat, close to one's body, where they can remain warm. Attaching hand warmers to electronic devices can also help.

Colder gunpowder does not expand as much as hotter gunpowder because its chemical reaction releases a fixed amount of energy into the system. When the propellant starts at a lower temperature, more of the energy is consumed bringing the gunpowder up to its burning point, leaving less energy for propulsion. As a result, **colder ammunition produces a lower muzzle velocity than hotter ammunition.**

This energy loss is only significant when a rifle is zeroed at a very hot temperature and then shot at a very cold temperature (or vice versa). Therefore, ammunition is ideally always shot at the temperatures in which zeroing occurred. This can be done either by zeroing in the same climate as one shoots, or by keeping warming up or cooling down the ammunition itself.

Radiation and Temperature Cycle

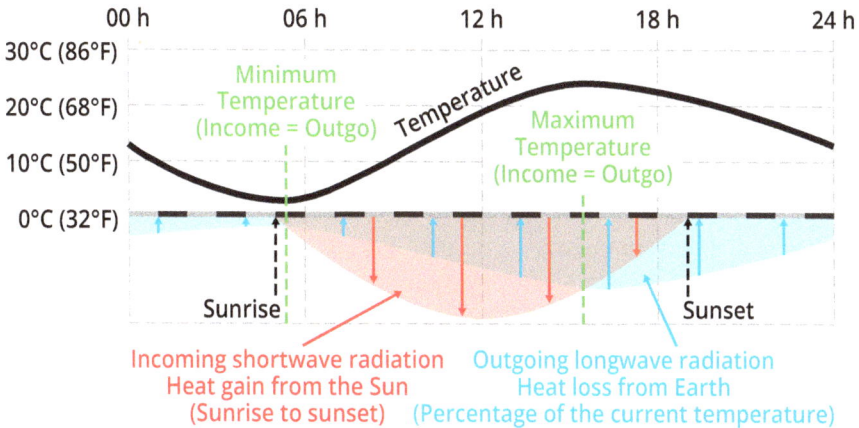

Image 59: Over the course of the day, **the temperature can significantly change** depending on the location and climate. In the example, the temperature ranges from 2°C (35.6°F) just after sunrise to 25°C (77°F) at around 16:00 (4 PM). Water retains heat and so smooths out daily temperature changes. Therefore, **inland, mountainous, and desert regions are prone to large temperature swings**.

Beyond equipment degradation, **cold temperatures also greatly affect the marksman's ability to shoot**. A cold body temperature leads to body tremors and shaking that make aiming difficult. Cold fingers go numb, reducing the feedback a marksman receives from pulling the trigger. The pain of cold is just generally distracting and can reduce focus and concentration.

Therefore, proper attire is vital for comfort and performance in cold weather. Dressing in layers that offer wind resistance and insulation helps maintain core body temperature. The human body constricts blood vessels in the limbs when the core is cold, so **warming the core is the best way to warm up one's arms and fingers.** However, the most effective way to "warm" hands is to not let them get cold. Utilizing hybrid glove-mitten combinations with hand warmers allows for both warmth and functionality. Whatever attire is used, a marksman must practice and dry-fire in that specific attire to find any significant interactions between their attire and equipment.

Insulating the body from the ground is also critical, especially when shooting from a prone position on snow or ice, which can rapidly drain body heat. A proper shooting mat, or even some spare pieces of cardboard, can serve as an effective barrier against frozen ground.

Dehydration is a common occurrence in cold weather because people dislike drinking very cold water when they are already cold. This a serious problem, as serious dehydration for extended periods can cause performance issues on par with being legally drunk. To address this, marksmen can warm up their liquids either with external heating or by keeping containers inside their coat to let their body heat up the liquid.

Managing condensation from breathing is important to prevent fogging or frost on optics. Exhaled breath must be directed away from equipment, and adequate air circulation must be allowed for. Historically, some snipers have reportedly put snow in their mouths to reduce the visibility of their exhale, but this is excessive in a modern conflict where isolated sniper teams operate at significantly farther distances. Instead, marksmen can simply breath through the nose, which directs air down and not forward.

Finally, **cold weather affects the environment**. For example, cold air is denser than warm air, increasing aerodynamic drag on bullets and causing them to lose velocity more rapidly. This loss of velocity can result in lower points of impact over long distances. Accounting for this is as simple as inputting the temperature into a ballistic calculator.

After shooting in cold weather, a common issue is the formation of condensation on the rifle when it is brought indoors and begins to warm up. Condensation occurs when warm air quickly cools upon contacting the cold rifle, causing excess water vapor to be released and condensed into liquid water. This water condensation can rust a rifle's metal parts. To avoid condensation on the rifle's surface, marksmen can wrap their rifles in towels which both absorb excess water and insulate them to slow the warming process and prevent condensation from forming in the first place.

4.e Shooting through Barriers

On rare occasions, a long-range marksman may want to hit something on the other side of a barrier. The most rigorously studied example is shooting through glass, so that is the primary subject of this section. That said, the conclusions would equally apply to other situations, such as shooting an animal hiding behind a thick bramble patch.

How well a bullet penetrates a barrier depends on two things: the energy it carries and its robustness. **A bullet that has more energy and is more robust is better able to penetrate a barrier and accurately impact a target on the other side.**

The energy contained in a bullet is proportional to the mass, but proportional to the square of the velocity. Therefore, faster bullets are more likely to penetrate and less likely to be deflected. This can be somewhat misleading though, as the mass of a bullet somewhat limits its potential

Shooting through Barriers

Image 60: An Italian soldier shoots through glass with a SIG Sauer SSg2000. **Non-laminated glass shatters**, so a second marksman can then shoot through the opening. Hohenfels, Germany, 15 Jun 2015.

Image 61: Snipers engage 100m-targets through glass panels. Without prebreaking the glass, **the target must be relatively close behind the glass**. Hohenfels, Germany, 20 May 2017.

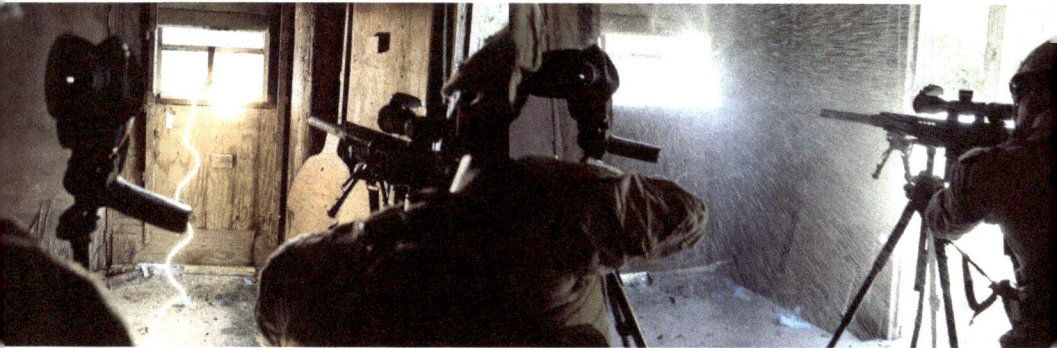

Image 62: A Marine explodes a **water charge** with a detonating (det) cord to clear the glass from a window. Camp Lejeune, NC, 16 Jul 2013.

Image 63: Once the glass shatters, it can send fragments backwards, so **eye protection is always necessary**. Camp Lejeune, NC, 16 Jul 2013.

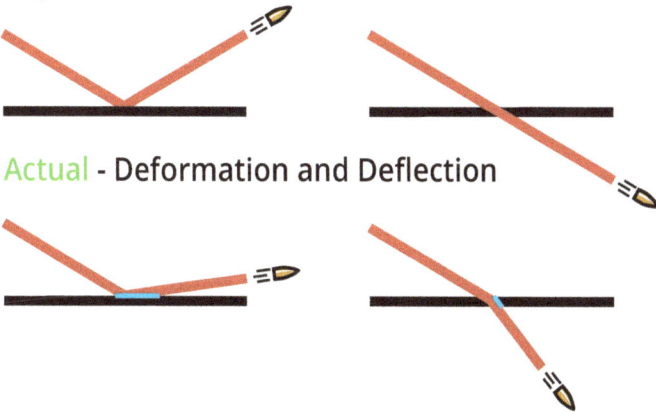

Impossible - Perfect Reflection or Perforation

Actual - Deformation and Deflection

Image 64: Bullets do not travel in a straight line through barriers.

velocity; therefore, mass and velocity typically increase in tandem. Similarly, the diameter, or caliber, of a bullet correlates well with its mass. Therefore, marksmen and manufacturers often use the caliber as a proxy measurement for a bullet's energy, even though it is not a particularly good proxy. That said, while specifics depend heavily on the specific bullets compared, in general, a .30-caliber bullet is much more likely to accurately penetrate a barrier than a .22-caliber bullet is.

A much better measurement of a bullet's energy is to calculate the actual energy by measuring the bullet's velocity at the point of impact using a ballistic calculator and multiplying it by the bullet's mass. Alternatively, a marksman could compare the grains of powder each bullet uses, although this does not account for the energy that is lost to air resistance during the bullet's flight.

The material composition of the bullet also affects its penetrative ability. Every type of bullet is designed to exist on a spectrum of over or under-penetration. Obviously, armor-penetrating bullets perform the best at penetrating a barrier. (In fact, they may not stop after penetrating the target, so knowing what is behind the target is critical.) On the opposite end of the spectrum, fragmentation bullets are designed to fragment immediately upon impact. While they may technically break through a barrier and exit the other side, the "bullet" would be nothing more than a spray of (still dangerous) fragments. Between those two extremes, hollow-point bullets fragment a lot, full-metal-jacket bullets fragment somewhat, and bonded bullets fragment a little.

The most significant factor when shooting at a smooth surface is the angle of that surface. **Bullets deflect off surfaces that are too angled.** For glass, shots are best taken if the surface is less than 15° from perpendicular. Anything higher than 15° requires extreme power from high-powered cartridges. For example, a .30-caliber bullet can usually shoot through glass at a much greater angle than a .22-caliber bullet can. Regardless, shots at window-glass at more than 45° from perpendicular are always too unreliable to be taken.

When a shot is taken at (non-laminated) glass or other material, the bullet exits the other side alongside fragments (a.k.a. "spall"). (See Image 60, Pg. 55.)This spall ranges from the size of dust to the size of the bullet itself and even larger. **The danger of spall is that it may cause collateral damage**, so it is important to consider the surroundings of any target behind a barrier. Even small wood chips and glass dust can be extremely hazardous to the eyes of anyone in the area. Maiming a deer or a hostage is unacceptable.

The exact exit of a bullet shot through a barrier is not predictable; instead, marksmen predict that **the bullet exits somewhere within a cone**. This cone has been the most well-studied for glass. For a bullet shot at a perpendicular pane of glass, the cone of the possible bullet trajectories begins at the glass and expands on the other side at a rate of about 0.5 to 2 cm (0.3 to 0.8 in) in diameter per 30 cm (12 in) of distance traveled. The cone of all possible fragments and glass dust is significantly wider, at about 10 cm (4 in) in diameter per 30 cm (12 in). When bullets pass through an angled barrier, they are deflected within a cone that is more perpendicular to the barrier, causing them to deviate from a straight trajectory. (See Image 64, Pg. 55.)Consequently, when shooting through a barrier, it is ideal to have a target be as close to the barrier as possible, and shoot as perpendicular to the glass as possible. (See Image 61, Pg. 55.)

Of course, the range of possible trajectories is only so wide because it must be comprehensive of all of the possible bullets, glass types, and glass thicknesses. When a marksman plans on shooting with a specific setup and through a specific barrier (including other material types), they can get far more precise predictions. For example, for every 30 cm (12 in) that a bullet may travel, it may deflect in a particular direction by 1 cm (.04 in), give or take ⅓ cm (.012 in) (i.e., the standard deviation may be only one-third of the mean prediction).

The most effective way to shoot "through" a barrier is to remove the barrier entirely. In fact, this is the only method that works when the target is far behind the barrier. In the case of glass, there are two common techniques for immediately removing the barrier. The first is to have two marksmen fire in quick succession; the first shot breaks the glass, removing the vast majority of the resistance for the second shot. The second technique is to attach an explosive, such as a blasting cap, to the glass and fire the blasting cap just before firing the rifle. (See Image 63, Pg. 55.) For either technique, eye protection is a necessity.

The second most effective method is to practice with the exact material one plans on shooting through. For example, elite military units may purchase and shoot the exact type of vehicle they plan to shoot through in an actual mission. And hunters may test the ability of their ammunition to shoot through brush and bramble while staying intact.

Advanced Effects

Advanced Effects, 600 Meters and Beyond

Problems worthy of attack prove their worth by fighting back.
—Piet Hein, Danish scientist and mathematician

A natural cutoff in shooting equipment occurs at various distances. For example, at about 300 m or yd, the naked human eye can no longer distinguish targets, making high-powered scopes necessary. A similar cutoff exists at 600 m or yd, where **tight part-tolerances** (i.e., low empty space between parts) become necessary to shoot accurately. These tight-tolerance parts are rigidly held in place by the other parts, and so enable more consistent shooting.

The reason loose-tolerance parts are used before 600 m or yd is because they make rifles more durable. The space between parts is where dirt and grime can be shoved when such a rifle fires, allowing that dirty weapon to continue to be mostly accurate. In contrast, a tight-tolerance weapon without that extra space available, may sustain internal damage upon firing as any excess dirt (especially quartz sand) is pushed through the chamber and bore. This principle explains why the maximum effective range of the standard-issue U.S. Army M4 is designed to only be 600 m. If the range were any longer, the rifle would be less reliable in harsh wartime environments.

There is another functional range-cutoff at 1000 m or yd, which is why shooting at that distance is called extended or **extreme long range shooting** (ELR). First and foremost, the cutoff is at "1000" because that is a nice round number. However more practically, this is the rough distance at which many **additional effects begin to significantly affect the trajectories** of bullets. These effects include parallel winds, transitioning to subsonic bullet speeds, spin drift, and the Coriolis effects.

5. Trajectory Information

As bullets travel farther distances, they must travel in more arced trajectories to hit their targets. In addition, there are more environmental factors that can alter a bullet's trajectory as it travels. Therefore, it becomes more difficult to predict where a bullet goes as it travels farther from the marksman. For a marksman to be as accurate as possible then, getting feedback on how a bullet travels after being fired is critical. Specifically, a marksman can use various methods to determine how a bullet travels, where it impacts, and how to incorporate that information into both their following shots and their ballistic calculators.

5.a Determining Bullet Trajectory

An observer may be able to see the location of a bullet as it travels using one of four methods: glint, tracer rounds, bullet trace (a.k.a. bullet wake), and vapor trail. (See Image 65, Pg. 61.) (Bullet trace is unrelated to "tracers," which are phosphorus bullets that light up.)

Glint is the most straightforward of the four; it is a reflection of sunshine on the bullet itself. Glint happens in rare instances when the sun shines on the bullet but not the surrounding area. For example, a 20-degree sun at sunset often may reflect off a bullet's copper jacket, producing a small, bright reflection that is visible to the naked eye. However, glint is not very helpful because it is so small and usually only appears in the part of the trajectory for which the sun is not blocked.

A "**tracer round**" is a type of ammunition that contains a small pyrotechnic charge that burns after being fired to create a visible trail of chemically produced light along the bullet's entire trajectory. This bright light allows anyone and everyone to easily track the bullet's trajectory. Modern marksmen never use these rounds because the enemy can see the origin of the tracer (i.e., the marksman's position), endangering the marksman to counterattack. However, historical marksmen would use tracer rounds to point out targets for other, stronger weapons such as tanks. And modern machine gunners use tracer rounds to see where their bullets are landing, since otherwise they may not be able to know.

In contrast to glint and tracer rounds, bullet trace and vapor tail are very useful indicators. They both appear for the entire length of a bullet's trajectory (under the right climate conditions) and do not give away nearly as much information to potential enemies.

Specifically, **bullet trace** is a form of mirage unrelated to tracer rounds. Mirage is a pattern of distorted light that occurs when light passes through pockets of air with different densities, bending the light in random directions. (How light travels through air depends on the air's density.) Thereby, bullet trace, like any other mirage, appears as a wavy pattern in the air.

Bullets create this wave of pressure and mirage in a cone shape as they travel and push air out of their way at supersonic speeds. This change in pressure, however, only changes the refractive index of the air (i.e., how air

Image 65: An F-35A Lightning II demonstration-team pilot and commander takes off. This jet shows "**trace**" (in the form of mirage), **vapor trail,** and **glint** in the same image. The mirage is created from the heated air coming out of the jet engine. While the mirage in bullet-trace is caused by mechanical force, it looks the same as this mirage. The vapor trail is formed at the tips of each wing and appears as white streaks or clouds. The glint appears on the top of the cockpit and at the top of the right wing as a shiny reflection. Hill Air Force Base, UT, 12 Feb 2020.

distorts light) very little, and so light has to travel through a lot of mirage for the bullet trace to be visible to a marksman.

Therefore, **bullet trace is best seen from directly behind the marksman** where the observer's line-of-sight aligns the most with the bullet's trajectory. (See Image 66, Pg. 63.) (See Image 67, Pg. 63.) Bullet trace is often impossible to see from even a small angle away from the bullet's trajectory because the amount of distorted air that light travels through decreases as the line-of-sight deviates from the trajectory of the bullet

There is no trace in many climate conditions. For example, wind erases trace. In fact, wind can sometimes push trace making it less reliable. High humidity and direct sunlight tend to create a more visible trace. Trace is easiest to see when the sun is not in front to minimize the amount of direct sunlight that can blind an observer. If the background is land and not air, trace is easier to see against a stationary, solid-color backdrop. When the marksman controls the targets, using stationary, solid-color targets helps minimize distractions and makes traces easier to detect.

Because trace is subtle, high-quality optical equipment is necessary; in fact, **marksmen often buy expensive optics specifically to see trace**. Similarly, shooting faster, larger, and generally less aerodynamic bullets causes a larger surrounding pressure wave and thereby a stronger bullet trace. In contrast, aerodynamic bullets are designed to minimize air resistance, and thereby the pressure wave. Therefore, marksmen must be aware that upgrading to a bullet with a higher ballistic coefficient may make trace more difficult to see.

If a spotter is looking for trace, they sit behind the marksman's rifle barrel by a few meters and align themself to the orientation of the rifle barrel as closely as they can. (See Image 66, Pg. 63.) That is, the spotter can be as far behind the marksman as they want. While sometimes a spotter may lay on top of a marksman to align themself as closely as possible to the marksman, this is only for concealment; simply sitting a short distance behind the marksman is sufficient to see trace and easily communicate. (See Image 67, Pg. 63.)

The best way to view trace is with a **low-magnification setting** to see a larger portion of trajectory of the bullet (usually between one third and two thirds of the total trajectory). To maximize the amount of trace seen, an observer can focus on a bullet's breakpoint. Subjectively, this is the point in the bullet's trajectory where the bullet begins to experience significant bullet-drop (i.e., just before the apex of the bullet's arc). Objectively, the breakpoint is the endpoint of the longest line that intersects the tunnel of air surrounding the bullet's trajectory. Focusing on the breakpoint ensures that the observer sees a large fraction of the trajectory where trace is clearly visible as close to the target as possible.

Seeing **the breakpoint and the target in one sight picture** is especially useful for determining whether a bullet is deviating too far to the right or left. In contrast, bullet impact is most useful for determining whether a bullet is impacting forwards or backwards. (See Determining Bullet Impact, Pg. 64.)

After positioning their optic, the marksman or spotter then relaxes their eyes and focuses their attention above the target, where a bullet would arc in the air. Bullet trace appears and disappears quickly, so the easiest way to see it is to have the trace appear where the marksman is already looking. Using a camera to record the trace is also helpful. Having a recording also helps an observer to determine the exact location of the trace of future bullets. It also avoids the problem of flinching and losing sight when the bullet is fired.

The fourth and final visual indicator is **vapor trails**. Although sometimes confused with bullet trace, a vapor trail is a rare and separate phenomenon. (Even rarer is a frost trail, which is a frozen vapor trail.) A vapor trail forms when a bullet displaces air, creating a low-pressure pocket behind it. In this

Spotter Position to see Bullet Trace

Image 66: 3rd Special Forces Group Soldiers engage targets with the M2010 rifle. To intersect the most with a bullet's trajectory, and therefore see the thickest trace, the spotter must be **vertically in line with the rifle**. Nellis Air Force Base, NV, 27 Aug 2019.

Image 67: Two infantrymen from the 101st Airborne Division conduct sniper training. In tactical settings, it is not uncommon to have the spotter sit or **lay on the marksman** to allow for non-verbal communication. Mielec, Poland, 09 Aug 2022.

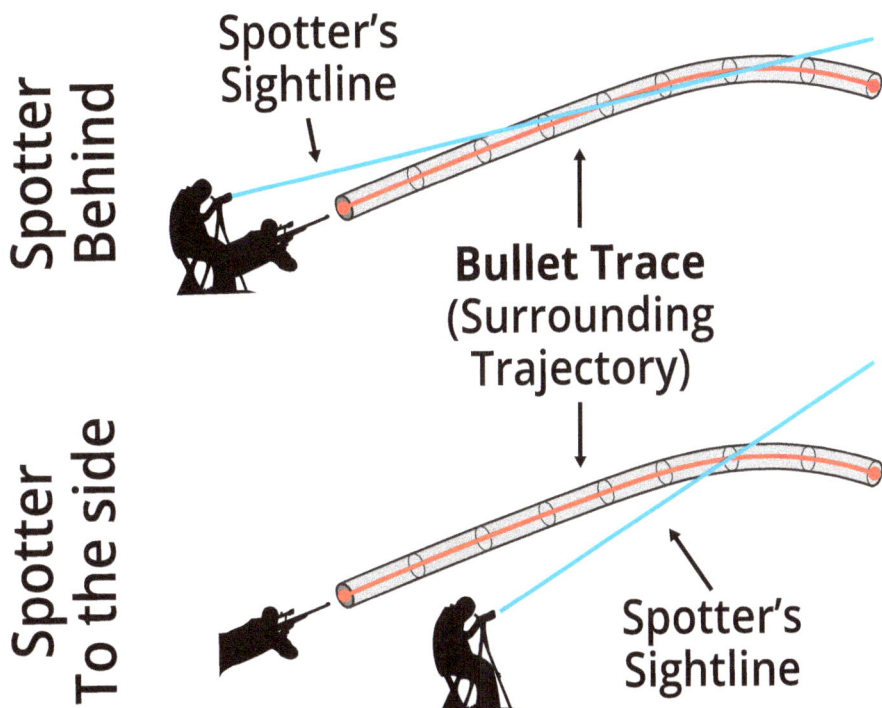

Spotter Behind

Spotter's Sightline

Bullet Trace (Surrounding Trajectory)

Spotter To the side

Spotter's Sightline

Image 68: Bullet trace appears as a rough cone that travels behind a bullet, which is caused by the bullet's supersonic pressure wave. (See Image 95, Pg. 95.) The cone follows the bullet's trajectory, forming a bent cylindrical path over time. The spotter **must intersect their sightline with that bent cylinder** to see the thickest section of the bullet trace possible. Therefore, they position themself behind the marksman's boreline and aim their sightline at the breakpoint of the trajectory. Even a position slightly to the side significantly reduces the intersection.

pocket, water vapor condenses into a **visible white cloud** because lower-pressure air cannot hold as much moisture. Consequently, vapor trails require very high humidity to develop. The resulting cloud can appear as a long line or as a dot trailing the bullet, since the water vapor can be quickly reabsorbed once the pressure returns to normal. Additionally, strong winds can prevent a vapor trail from forming altogether. Overall, the climate conditions necessary for vapor trails are very specific, and these trails can disappear as quickly as they appear.

Unlike glint or trace, a vapor trail following the bullet can be seen from any direction. In fact, since vapor trails appear as white dots or streaks, they do not require special skill, positioning, or training to see. That said, often the white dot is too small or moving too fast to be particularly useful to a marksman. They are also so rare (or at least climate-specific) that it is difficult to practice with them. Many marksmen never see a vapor trail because their bullets are too aerodynamic, small, or slow to form a low enough pressure behind them. Or they simply may never shoot in cold, high humidity weather.

Of course, there is debatably a fifth way of determining a bullet's trajectory, and that is if the bullet **hits something along the way**. When shooting in an area where overhead structures such as bridges or trees are present, there is always a risk of unintentionally hitting something. Therefore, marksmen must always consider and avoid the possibility that the highest point in a bullet's trajectory (i.e., the "maximum ordinate") intersects with some object. This can be an issue when using a scope at a high magnification. In that case, a marksman may only see their target and forget to look for potential obstacles above the sight picture, along the bullet's trajectory.

5.b Determining Bullet Impact

For long-range marksmen, missing some shots is inevitable. Therefore, a marksman must develop the skill of knowing where they missed so they can implement corrections quickly and accurately.

The best method for determining the location of a bullet's impact while shooting is to pay attention to the **area surrounding the target**. A bullet typically shows clear signs of where it impacts, which can be seen in the surrounding area. For example, when a bullet hits dirt, the impact typically throws a **characteristic cloud of dirt into the air.** (See Image 69, Pg. 65.) For any dust cloud, the lowest point of that cloud is the most likely impact point, since the cloud is always above the impact site.

Sometimes, dirt clouds can be very difficult to see. Therefore, one trick that many marksmen use is to upgrade to more powerful ammunition. In

Image 69: The dust cloud to the right is caused by a bullet impacting the berm. Because the berm is sloped, dust explodes upwards. Therefore, a marksman knows that **the bullet impacted at the base of the dust cloud**. After the dust enters the air, a marksman can use the dust to determine the prevailing winds. North West Islands, South Korea, 10 Oct 2017.

fact, one of the primary reasons that long-range marksmen use such powerful ammunition is not to get the bullet to the target (less powerful bullets can often travel sufficiently far and cost less), but instead to create larger dirt clouds when the bullet misses. Thereby, the marksman can more easily correct their shots. Another solution to better see dirt clouds is to set up cameras at the target site to see impacts where they happen.

It is important to note that visible signs of dirt displacement do not always indicate a missed shot. For example, when shooting at steel targets, the bullet fragments upon contact. Individual fragments then impact the ground and can cause dirt to kick up in a line across the target's base.

It can be especially difficult to determine whether a shot missed high or low. This is because depth perception does not work well at long distances. Many marksmen at longer ranges **set up their targets on the ground**, and do not elevate their targets like a closer-range marksman would. Thereby, if there is no sign of an impact, the impact could have likely occurred from a bullet falling over and behind the target, hiding the impact from observation. In sum, for a target set up on the ground, any impact below the target must be in front of it, and the bullet went too low.

In contrast, if the target is mounted high off the ground, a bullet may arc over the target and still allow the marksman to see the impact below the target's stilts. With the arc, the bullet would have landed behind the target.

However, because the bullet would be seen below the target, it would create the illusion that the bullet impacted in front of the target instead.

Because ground impacts are not always reliable or visible, marksmen must also look for additional clues. At close range, one sign of a missed shot would be a lack of a hole in a paper target. However, at longer ranges, marksmen use steel targets because they are durable and reusable, and also because they make a clear ring when they are hit. The **absence of that ring** would indicate a missed shot. That said, a lack of something (i.e., a hole or a ring) is not very specific to where the miss was.

When hunters and military shoot living targets, determining whether or not a shot hit or missed can be equally difficult. The response of a deer to a gunshot varies based on several factors, including the location of the impact on the body, the caliber and type of the bullet, the range and the angle of the shot, and the overall health status of the deer.

Observing the **immediate reactions of the deer** can provide insight into where the bullet has struck, thus guiding the subsequent follow-up actions to ensure the deer is dispatched humanely and swiftly. For example, when a deer collapses immediately and exhibits jerky or twitchy movements, it suggests that the bullet hit the key neurological areas of the head, high neck, or spine. If a deer falls immediately but attempts to sit up, it could be an indication that the bullet has damaged the spinal column at the saddle or upper haunch level. If the deer moves a short distance before collapsing, possibly in an uncoordinated manner or lunging forward from its back legs, they could have a chest, lungs, or heart shot. If the deer kicks out its back legs, stands with its body hunched, or walks off slowly with a stilted gait, it might signify a shot to the stomach, intestines, or liver. In this situation, the deer, if left undisturbed, might lie down after a brief period. Finally, a deer that initially collapses but immediately gets up and runs off, possibly indicates that the bullet hit the mid haunch, back leg, front leg, chest, outer neck, or head. These areas, while still painful, might not cause immediate incapacitation, allowing the deer to regain its footing and attempt to flee.

Enemy targets may exhibit all of the above symptoms as well. In addition, clothes may show visible blood splatter and staining. People are often near clean surfaces, so blood surrounding the target is another indication of a successful shot. Humans are also not as prone to running, and may collapse in an unnatural way as if they have lost their structural integrity.

Because there are so many different, unpredictable responses, marksmen must ensure that they **perform a proper follow-through** by reengaging their sight picture. Reengaging allows a marksman to remember and internalize where that target was when the shot was taken. In contrast, the most common response to shooting a deer is to immediately want to chase after the deer, immediately forgetting the location at which the deer was shot! This makes finding the start of the blood trail much more difficult.

5.c Making Adjustments

After a miss, marksmen must readjust their point-of-aim according to a series of steps. The first step is to ensure that the miss was not due to faulty equipment. This is rare, but important. The most common equipment fault is **a scope that is loose or has lost its zero** due to an impact or recoil. Minor equipment failures are extremely difficult to diagnose while shooting, but the marksman must always be aware of that possibility instead of mindlessly trying to adjust their point-of-aim for hours.

Second, the marksman must ensure that they themself are not creating errors. For example, when using a ballistic calculator, it is very easy to input the wrong information. Therefore, checking and double checking inputs can prevent incorrect outputs.

Regarding technique, the marksman must ensure that they are shooting consistently from shot to shot. Specifically, marksmen must know their body mechanics: stance, grip, breath control, and trigger squeeze. To become more consistent, marksmen get in the habit of calling their shots specifically to remember their technique at the moment they shot and to reflect on any possible errors or deviations they may have made at that time.

Third, the marksman must **update the information available to them**. This update mostly involves rereading the wind and reverifying the distance to the target to improve the windage hold and elevation hold respectively. Similarly, climate data changes throughout the day, so marksmen must regularly update their ballistic calculator every hour or so.

Even if the marksman is confident in their windage and elevation holds, if a shot is off to the left or right, something was incorrect; and if it wasn't the equipment, and it wasn't the data, it was likely the information. The **marksman listens to the bullet, the bullet doesn't listen to the marksman**.

If an error was found and an adjustment must be made, the marksman **must correct their information precisely**. While sometimes using educated guesses for an adjustment is the only option, even then, using and recording actual numbers allows a marksman to shoot better with future shots. For example, a marksman can better reference previous shots when they were recorded as "2.2 mils" as opposed to "kind of leftish."

When shooting at a target, any adjustments must be greater than the width of the target. Requiring a minimum width for adjustments is useful because **marksmen as a group habitually underestimate how much they miss by**. By setting the minimum adjustment to the target's width, marksmen are far less likely to miss twice on the same side. Marksmen are also never forced to overcorrect and miss on the other side of the target, since a distance equal to the target's width by definition cannot pass over the target. For example, if a shot impacts 50 cm (20 in) to the left of a 100 cm (39 in) target, then adjusting only 30 cm (12 in) (i.e., a small adjustment) misses again. Instead, the marksman must adjust by at least 100 cm (39 in) (i.e., the width of the target).

A larger adjustment also allows a marksman to **bracket their estimates**. For example, if a shot is overcorrected and misses to the opposite side of the target, then the marksman knows that the proper hold is somewhere between their first and second shot. For example, if a windage hold of 2.0 mils (7.0 MOA) misses to the right and 2.4 mils (8.4 MOA) misses to the left, the marksman can be confident that a windage call of 2.2 mils (7.7 MOA) would hit the target in the center. In contrast, missing twice to the same side due to undercorrection doesn't help the marksman much.

If all else fails, the marksman might consider that a random factor, such as a sudden gust of wind or a corroded cartridge, caused the miss. In this case, there's no immediate solution other than to try again. If another error occurs, it clearly indicates that something within the marksman's control is wrong. However, if no further errors happen, the issue is resolved.

Nonetheless, attempting to shoot again without making any changes is a last resort because it essentially means giving up. Sometimes, recognizing that nothing can be done is the most efficient course of action. However, each time this thought arises, the marksman must ensure they are putting maximum effort and skill into every shot. While dismissing errors as unavoidable might feel reassuring in the short term, it ultimately undermines skill development and progress.

5.d Trajectory Validation for Ballistic Calculators (Truing)

Ballistic calculators predict bullet trajectories using mathematical models that place a curve onto the bullet's zero point to represent its prediction of a bullet's trajectory. These calculators make their predictions by assuming average, ideal conditions. While these predicted trajectories can be highly accurate,

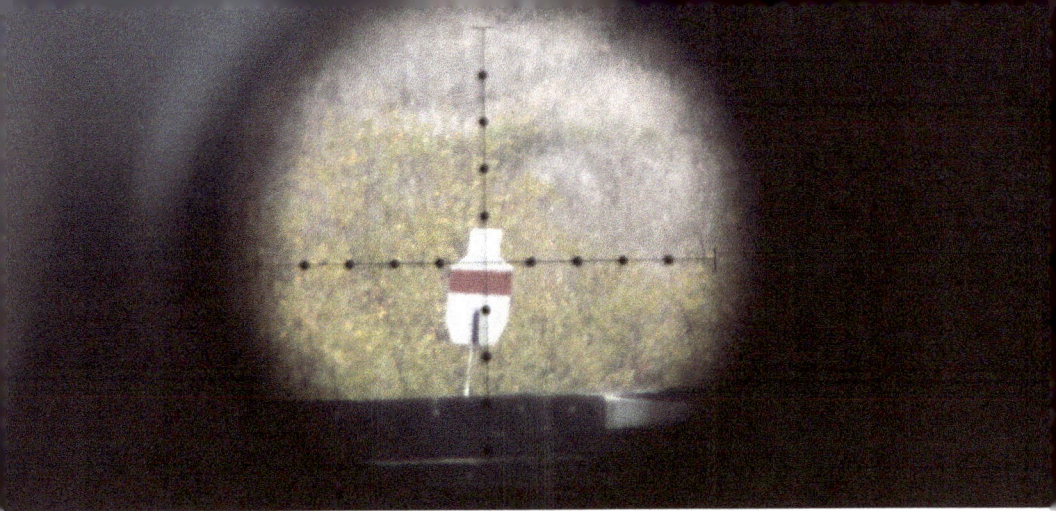

Truing

Image 70: This reticle is a Mildot reticle, so each dot is 1 mil (3.5 MOA) apart. The target is very close to 2.5 mils (8.75 MOA) tall. It is an E-Type target, all of which are 101.6 cm (40 in) tall or about 1 m tall. Therefore, it can be determined that the target is about 400 m (437 yd) away from the marksman. The red line (i.e., the truing line) is 0.5 mils or 20.3 cm (1.75 MOA or 8 in) tall. If the **truing line is hit using the trajectory predicted by the ballistic calculator**, then that predicted trajectory must be accurate within 0.5 mils. That is still a relatively large amount, and most precision long-range marksmen would use a narrower line or a farther placement. Camp Pendleton, CA, 30 Jan 2014.

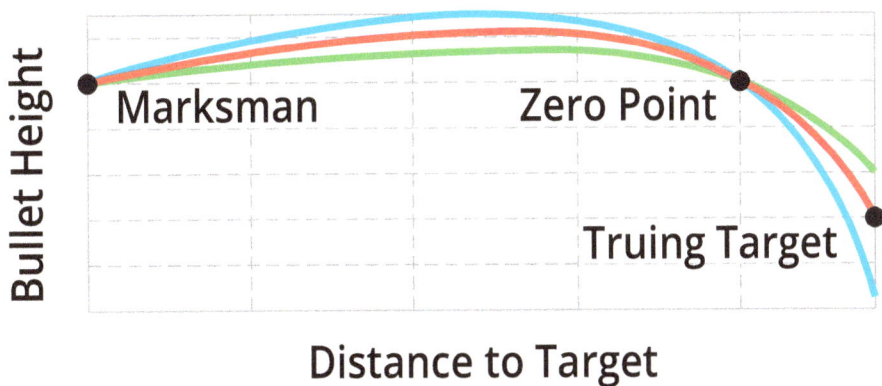

Image 71: A zero point is where the bullet's point-of-aim matches its point of impact. Because only two points (the marksman's position and the zero point) are not enough to define a single curve, the bullet could theoretically follow infinite trajectories between them. **"Truing" adds a third reference point** to solve this problem. With three points, a ballistic calculator can produce a much more accurate and predictive trajectory curve. In fact, a marksman can repeat the truing process at multiple distances, adding as many points as needed to achieve the most precise curve possible.

real-world factors, such as non-ideal weather or different ammunition, can lead to **discrepancies that shift the actual trajectory forward or backward** from the calculator's predicted trajectory. For example, if a properly zeroed bullet travels faster than the model predicts, it would drop less and hit higher beyond the zero distance (i.e., a forward shifted trajectory). Conversely, if the bullet travels slower than expected, it would land lower than the calculator's projection (i.e., a backward shifted trajectory).

To ensure accurate modeling, marksmen must "true" their system. **"Truing" is to shift a ballistic calculator's predicted trajectory so that it matches the actual trajectory of the bullets a marksman is firing.** This matching is accomplished by giving the ballistic calculator an additional point of data: the truing distance.

The truing distance is essentially a second zero distance. For both truing and zeroing, the first step is to determine how far the marksman's point-of-aim is from the actual point-of-impact. For zeroing, this difference is dialed into the scope so that the point-of-aim intersects the point-of-impact. In contrast for truing, the vertical angular difference is input into the ballistic calculator as the true elevation hold without adjusting the scope.

Thereby, the ballistic calculator has three X-coordinates (i.e., the marksman's location, the zero distance, and the truing distance) and three Y-coordinates (i.e., zero bullet-drop, zero bullet-drop, and the bullet-drop at the truing distance) on which to base its curve. The program can generate a much more accurate predicted curve when it is basing the curve on two verified coordinates instead of just one (i.e., the zero point alone).

The truing process begins by setting up a truing target (a.k.a., a "truing bar"). Ideal truing targets are steel rectangles that are short, wide, and embedded in something. Ideal truing targets are steel because steel makes a loud ring when shot. They are short because they are meant to calibrate the elevation hold within a narrow vertical band. They are wide because the windage hold is not being calibrated, and so where the bullet impacts horizontally is irrelevant. They are embedded because the marksman is expected to miss (if they never missed, truing would be redundant), so being surrounded by a different material helps the marksman to identify whether their shot was high or low.

A good option is to embed two stacked railroad tracks in a dirt berm. When the bullet hits a track, it rings; but when it hits the berm it makes a dust cloud. (See Image 69, Pg. 65.) A much cheaper, single-use option is to paint a line onto a large cardboard target. (See Image 70, Pg. 69.) Then, each bullet hole can be tracked as a hit or a miss. However, without a metallic ring, truing on cardboard can become more difficult at farther distances. Therefore, the marksman may want to set up a remote camera that can live-transmit the area surrounding the truing target.

Because the ideal height of a truing target is an angular height (e.g., mils and MOA) and not linear height (e.g., cm and in), a truing bar at 1000 m or yd must be larger than a truing bar at 400 m or yd. A common truing target height is around ½ mil (1.7 MOA). Although that height may seem tall for a precision rifle, the truing target must be **at least as tall as the inherent dispersion or precision of the rifle system**. As a counterexample, if a target were only 5 cm (2 in) tall (⅛ mils, ½ MOA at 400 m or yd), and a rifle had 15 cm (6 in) of dispersion (⅜ mils, 1.5 MOA at 400 m or yd), then over half of perfectly aimed rounds would miss the target no matter what. On the other hand, the height of the truing target is the maximum accuracy of a trajectory that is derived from it; for example, a trajectory that is trued on a ½ mil (1.7 MOA) target can only be predictive up to a ½ mil (1.7 MOA) accuracy.

When deciding on how tall to make their truing target, a marksman can reference the advertised precision of their rifle and ammunition. For example, a truing target for an off-the-shelf hunting rifle would be much taller than a truing target for an expensive rifle with match-grade ammunition, which may advertise a dispersion of 0.2 mils (0.7 MOA). A rifle system can be trued on a target that is equal to its advertised precision.

Once the target is set up, **the marksman records the elevation hold** that is required to hit the target. To do this, the marksman first uses the elevation hold that their ballistic calculator predicts. If that shot hits, the predicted trajectory is fine. If it misses, the marksman adjusts their elevation hold up or down for their following shots. Once the correct elevation hold is found, the marksman inputs that hold into their calculator so that the calculator can update its predicted trajectory to be either shorter or longer.

Whether the shot is high or low on the target is usually caused by ammunition that respectively travels faster or slower compared to average ammunition. Therefore, **some marksmen use a chronograph** to directly measure their bullets' velocities and feed the average velocities into their ballistic calculators instead of elevation holds. Traditionally, this has been an inferior method because using velocity alone may not account for other environmental causes of high or low shots. However, as ballistic calculators get better at integrating data and designing models for specific bullets, inputting actual velocity is more and more becoming a viable way to true a ballistic calculator's trajectory.

For example, older ballistic calculators had a limited set of trajectories to start truing from. Because these models were highly generalized, they didn't predict the trajectory of any particular kind of bullet particularly well. Many only had two basic trajectory models named "G1" and "G7" (clarification: these are drag curves, and trajectories are based on drag curves). However, modern calculators use hundreds of custom curve models, which are predictive models customized to each type of round that has ever been widely manufactured.

Were a marksman to use the older G1 or G7 models and also want to true those models with a chronograph, the resulting trajectories would not be accurate everywhere since the G1 and G7 models wouldn't have the same drag curve (i.e., aerodynamic drag as a function of bullet velocity). This means that while the predicted trajectory could be accurate for a part of its range, it would always be inaccurate for other ranges.

Some calculators can accept **multiple holds or velocities** at different distances so that a marksman can correct (i.e., true) their model every few hundred meters. In this way the marksman can essentially transform any model into a custom curve model for their ammunition. This can be somewhat redundant to the custom curve models already designed by the ballistic calculator manufactures. However, truing at multiple distances becomes more important at extremely far distances (e.g., over 2 km or 1.5 mi) because errors in predicted values accrue exponentially with distance. And truing at multiple distances is the only way to precisely fit a generalized trajectory (e.g., G1 or G7) to a particular rifle system.

6. Bullet Spin

Bullets are cylinders in the back with a pointy cone facing into the wind. This shape minimizes the air resistance the bullet faces, as the tip slices through an incoming airflow. However, this shape also makes a bullet prone to randomly flipping over (i.e., "tumbling") because the shape has a center-of-mass that is behind its center-of-pressure (i.e., where the air acts on the bullet). This tumbling would make the bullet face more air resistance, and therefore slow down faster, than if it could maintain the tip facing forward.

There are two ways to prevent tumbling. First, many projectiles move the center-of-pressure behind the center-of-mass by adding fins to the rear of the projectile. Such projectiles include: arrows, missiles, and airplanes. The other way **to prevent tumbling is to spin a projectile very fast**. Bullets use this method. Interior grooves in a rifle barrel, called "rifling" forcefully spin

Barrel Twist

Image 72: This 105 mm L7 tank gun barrel (not a rifle barrel) was cut to show the **rifling** inside. Typical rifle barrels use far fewer grooves.

Image 73: **Bullets expand into the rifling** after they are fired (bottom) to both grip the rifling and to plug the barrel to stop gas from slipping past.

bullets as they travel forward through the bore. (See Image 72, Pg. 73.) Specifically, a bullet expands to fill into the rifling grooves, forcing it to follow the orientation of the grooves as it travels. (See Image 73, Pg. 73.)

Why exactly spinning causes the tip of a bullet to continue to point forward as it travels is explained in this section. It is not at all intuitive that spinning a bullet would somehow cause its tip to stay forward. Then, this section explains the side effects of spinning a bullet that cause the bullet to deviate from its initial trajectory, called "aerodynamic jump" and "spin drift."

6.a Barrel Twist

The first step in understanding spin is to understand how it is created. Inside every rifle barrel are grooves that span the length of the barrel while slightly rotating. (See Image 72, Pg. 73.) That is, the grooves are helical. The grooves (cutout section) and the lands (raised section) are together called "rifling" and are the origin of the word "rifle."

When a bullet is shot through a rifle barrel, the extreme force of the expanding gas causes the bullet to deform into the grooves, forming a barrier between the rifle chamber and the muzzle. Because the bullet expands into the grooves, the bullet follows the rotation of the grooves as it moves forward, causing the bullet to rotate as well. Therefore, the bullet makes exactly one spin for each length of distance that the rifling takes to make one rotation (a.k.a., one twist) in the bore. The bullet then continues to rotate at this rate of **one spin per each twist-distance** even after it has exited the bore. For example, if a rifling were to have one rotation in 25 cm (10 in), then a bullet shot through that rifling would continue to make one rotation every 25 cm (10 in).

In other words, bullets spin fast because they travel fast. And the twist rate locks in the relationship between spinning speed and traveling speed at the muzzle. (Notably, velocity degrades faster than spin, so the twist rate lowers as the bullet travels forward.) For example, if a bullet were to travel at 1000 m/s (1093 yd/s) with a 25 cm (10 in) twist, it would spin at 4000 rotations per second at the muzzle (1000 m ÷ 25 cm/rotation = 4000 rotations).

Twist rate is often expressed as a ratio in inches or millimeters, such as 1:10 (1 twist in 10 in) or 1:254 (1 twist in 254 mm). The unit of measure is usually inferred from the number, as twist rates for common firearms only exist in a small range. For example, almost all rifles sold in the United States are between 1 twist per 6 in to 1 per 20 in (i.e., 1:6 to 1:20). In contrast, metric twist rates have a larger ratio because millimeters are smaller than inches.

The twist rate has a nominal effect on muzzle velocity because some of the energy required to spin is taken from the energy propelling the bullet forward; however, the loss is less than 1% of the total. Therefore, changing the twist rate (and only the twist rate) does not significantly affect bullet velocity.

6.b Spin Stabilization (Precession)

Spinning is effective at maintaining an object's orientation because of a weird quirk of physics: when any object spins and an external force is applied to it, **that force acts at a right angle to the direction from which it was applied**, and not in the direct line of the applied force. (See Image 74, Pg. 75.) This is known as "gyroscopic precession," or "precession" for short.

A common example of this phenomenon is a spinning top. When a top is still, it falls over. But when it is spun, it stays upright. A bullet, like a top, resists changes to its orientation because of gyroscopic precession. When any force is applied to the side of a spinning bullet like trying to push a top over while it's spinning (in physics terms, the "overturning moment or torque"), it doesn't move in the direction one would expect (directly away from the push). Instead, the spinning bullet moves at a right angle to the applied force.

Much as tops wobble on uneven ground, bullets wobble in the air because they travel through air pockets of different densities. Therefore, bullets only point exactly forward on average, and do not necessarily point forward at any particular instant. To stay pointed forward on average, a statically stable bullet rotates (in physics terms, "precesses") in a circle around the exact direction of airflow. This precession is similar to how the handle of a spinning top slowly circles the vector of gravity. (See Image 80, Pg. 78.)

This kind of deflection is known as "**static stability**." Static stability causes the force trying to overturn the bullet (i.e., air resistance) to be deflected sideways so that the bullet maintains its near-forward orientation. The effectiveness of static stability is directly related to how fast the bullet is

Precession

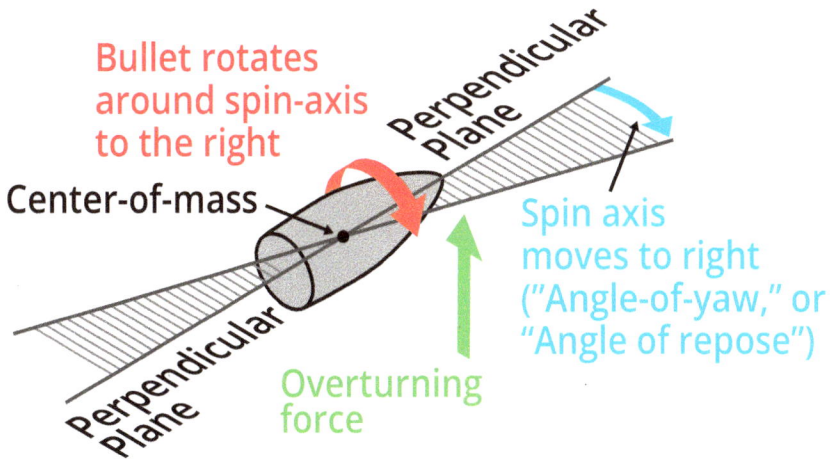

Bullet rotates around spin-axis to the right

Perpendicular Plane

Center-of-mass

Perpendicular Plane

Overturning force

Spin axis moves to right ("Angle-of-yaw," or "Angle of repose")

Image 74: Precession is a phenomena whereby when an overturning force (i.e., a force that causes a rotation perpendicular to the spin-axis) is applied to a spinning object, **the spin-axis of that object rotates 90 degrees** relative to the direction of applied force.

Yaw and pitch for a bullet relative to the direction-of-travel

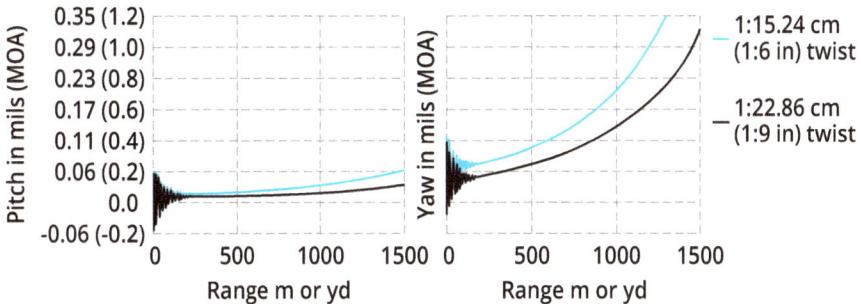

1:15.24 cm (1:6 in) twist

1:22.86 cm (1:9 in) twist

Image 75: The spin-axis of a flying bullet deviates from the bullet's direction-of-travel by a small vertical angle, called the "pitch angle," and a small horizontal angle, called the "yaw angle." This is caused by an overturning force (air resistance) being diverted through precession from changing the pitch to changing the yaw. **The greater the angular velocity of the bullet's rotation (i.e., faster spinning caused by a shorter twist), the more overturning force is diverted** to affecting the yaw instead of the pitch. Thereby the pitch cannot reorient as quickly. At the initial stage of flight, the bullet's tip rotates in a dampening spiral caused by the initial yaw angle formed when the bullet leaves the rifling (i.e., lateral throwoff). After approximately 200 m or yd, the angles stabilize due to dynamic stability.

Spin-axis Alignment

Image 76: **Bullets perform just like spinning tops**. That is, both reorient to align with the vector of force (gravity for tops, and air resistance for bullets) by facing resistance to their spinning (surface friction for tops, and air resistance for bullets).

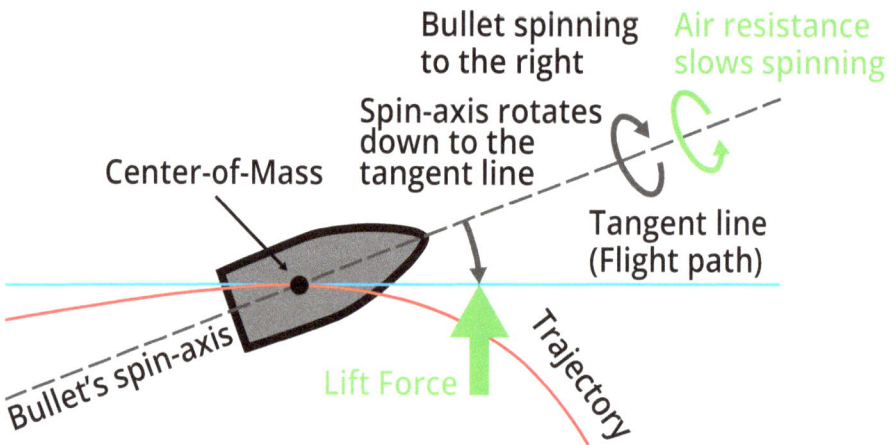

Bullet spinning to the right

Air resistance slows spinning

Spin-axis rotates down to the tangent line

Center-of-Mass

Tangent line (Flight path)

Bullet's spin-axis

Lift Force

Trajectory

Image 77: There's a balancing act between two types of friction on a spinning bullet. One is the lift force (See Image 83, Pg. 83.) that drives the bullet's precession around a circle (See Image 74, Pg. 75.); the other friction is the air friction that slows the bullet's spinning. The friction that slows the bullet's spin is what causes its spin-axis to align with the incoming airflow, whereas the friction resisting its precession causes the bullet to unalign.

spinning. However, the spinning speed cannot be increased indefinitely due to material limitations. For example, if a copper jacket were spun too fast, it would tear away from the rest of the bullet (i.e., mechanical failure due to centrifugal forces).

The **gyroscopic stability factor** (Sg) measures how fast a bullet must spin to maintain its orientation during flight (i.e., its static stability). By definition, an Sg < 1 means that a bullet is not stable and cannot maintain any orientation. In contrast, an Sg ≥ 1 means, by definition, a bullet is spinning fast enough to maintain its orientation.

The specific Sg of a bullet depends on the air's density and the bullet's density, length, and shape. As a rule though, longer bullets require faster spin rates to stay stable than shorter bullets do. (In physics terms, the farther apart the center-of-mass and center-of-pressure are, the faster a bullet has to spin to be stable in a given airflow.) Because bullet length increases the ballistic coefficient (BC) (i.e., how aerodynamic a bullet is) but decreases the static stability, more aerodynamic bullets usually need more spin to remain stable.

That in turn means long, narrow bullets must be fired with a higher twist rate or a faster velocity. Increasing the twist rate directly increases the spin, and is a simple solution to keeping long, narrow bullets stable. However, increasing the velocity both increases the amount of spin (for a given twist rate) and also increases the amount of spin required for stability. In sum, a 3% increase in velocity may only result in a 1% increase in spin rate.

When determining the optimal twist rate or velocity for any given bullet, marksmen must be careful to measure or determine the Sg at sea level (or at least at a lower altitude than they intend to fire). Because thicker air decreases Sg, what may be a sufficient Sg at a higher altitude may become insufficient when shooting at a lower altitude. (See Image 100, Pg. 99.) For example, because the stability factor scales inversely with air density, and air density at the top of Mount Everest is 34.8% of the air density at sea level, a bullet with an Sg of 2.87 (1 ÷ 0.348) on top of the mountain would only have an Sg of 1 at sea level. Oddly enough, because of this fact, **marksmen at lower altitudes must use rifles with lower twist ratios**.

Determining the optimal twist rate to achieve a desired spin factor for a specific bullet and climate involves complex formulas. Today, manufacturers typically perform these calculations due to their complexity. These calculations take into account various factors, including bullet mass distribution, shape, spin dynamics over time, spin rate, behavior in transonic and subsonic velocity ranges, velocity, and air density. Additionally, manufacturers have access to empirical data that is not available to the general public. They are

Helical Spinning

Straight (False)

Helix or Spiral (Stable True)

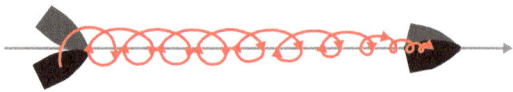

Tumbling (Unstable True)

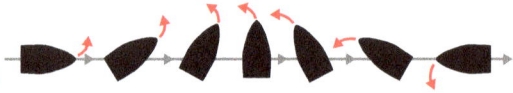

Image 78: In animation and movies, bullets are usually depicted as traveling in a straight line. Actually, **bullets revolve about the trajectory as they travel**.

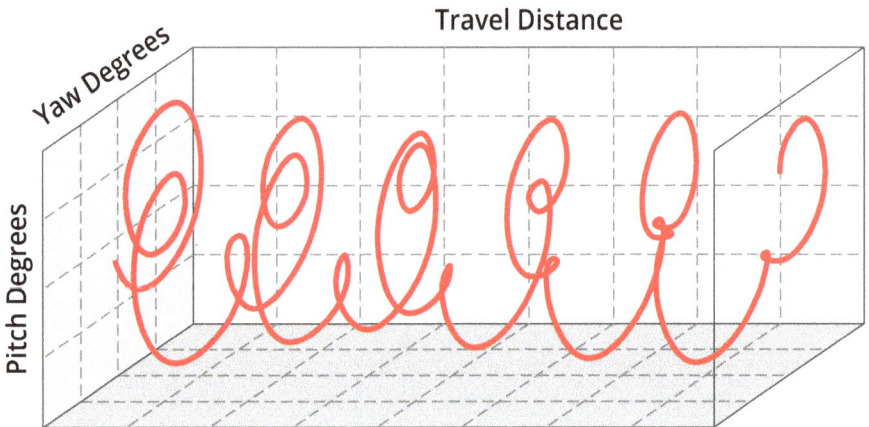

Travel Distance

Yaw Degrees

Pitch Degrees

Image 79: This image depicts the angle at which an actual example bullet pointed for a given segment of its travel. (See Coning and Nutation, Pg. 118.)

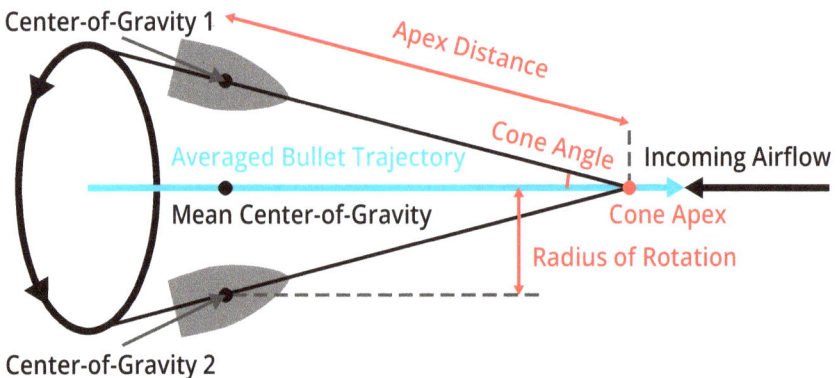

Center-of-Gravity 1

Apex Distance

Cone Angle

Incoming Airflow

Averaged Bullet Trajectory

Mean Center-of-Gravity

Cone Apex

Radius of Rotation

Center-of-Gravity 2

Image 80: A bullet is often depicted as spinning its tip more than its tail. The more likely configuration is that the tail rotates about a larger radius than the tip.

also better equipped to determine optimal twist rates for non-standard bullets, such as polymer-tipped or tungsten-cored varieties. As a result, conducting a simple online search usually provides more accurate information on suitable twist rates for different types of ammunition than any layman could hope to achieve on their own.

If a marksman already owns a rifle, they would need to research ammunition before purchasing it to ensure that it is compatible with their rifle's twist rate. If a marksman is buying a rifle before knowing what ammunition they would use in it, they are better off buying the lowest twist rate available so they can use the widest variety of ammunition. In contrast, if a marksman does know the ammunition they would use, they can look up the manufacturer-recommended twist rate for that ammunition. If the marksman plans to use multiple kinds of ammunition, they must choose the lowest of the various recommended twist rates.

To determine the twist rate of a rifle, a marksman begins by attaching enough cotton to the end of a cleaning rod to engage their rifle's rifling. The cleaning rod is then inserted a short distance into the bore. Using a marker, the marksman draws a line across both the rifle's crown and the cleaning rod, ensuring the lines are aligned. The rod is pushed downward until it completes one full rotation, causing the marked line on the rod to realign with the mark on the crown. The marksman marks the rod again at this point and withdraws it from the bore. The distance between the two markings on the cleaning rod represents the length of one full twist in the rifle's rifling.

6.c Bullet Orienting to Vector (Weathervaning, Tractability)

A statically stable object maintains the orientation of its spin-axis despite the application of an external force. A second kind of stability, **dynamic stability,** is the ability of an object to reorient its spin-axis to align with incoming force. The way that dynamic stability works is that the tip of a bullet precesses (i.e., rotates around the spin-axis) in increasingly small circles. In contrast, if a bullet is statically stable but not dynamically stable, then the bullet would precess in increasingly large circles until the Sg falls below 1 and the bullet tumbles. In other words, for a dynamically stable bullet, the angular difference (i.e., the angles of yaw and pitch) between the direction of incoming airflow (i.e., the bullet's direction-of-movement plus wind) and the orientation of its spin-axis becomes smaller with time.

Tops demonstrate dynamic stability by adjusting themselves to stand perfectly vertical, regardless of the initial direction of their spin-axis. Similarly, bullets reorient themselves to align with incoming airflow because they experience air resistance as an external force, much like how tops are affected by gravity.

The mechanism by which dynamic stability works is that the slowing of the spinning itself causes the bullet to reorient. (See *How Do Gyroscopes Lift Themselves Up?* from The Action Lab on Youtube.) Bullet's can continually lose spin while increasing their static stability at the same time because they lose velocity faster than they lose spin. A slower bullet faces less air resistance, and so can remain stable with less spin. In fact, a bullet's static stability at the point-of-impact can be more than 5, depending on how much a bullet has slowed down.

As a bullet travels in an arced trajectory, it goes up and then comes down. Therefore, the airflow it faces (i.e., the air resistance) comes from above and then from below, since airflow and air resistance occur in the opposite direction of a bullet's movement. A dynamically stable bullet always points its tip into incoming airflow because the spin-axis of a bullet goes through the bullet's tip. In that way, **a dynamically stable bullet is a weathervane**, pointing into incoming airflow. Because all modern bullets exit the muzzle with dynamic stability, bullets rotate their spin-axis forward as they travel through their trajectory.

Because the reorientation of the spin-axis is due to incoming airflow, crosswinds also affect where the spin-axis points. (See Aerodynamic (Muzzle) Jump, Pg. 86.) However, the effect of crosswinds on the spin-axis is minimal because wind speeds are typically much lower than bullet speeds. For example, if a 16 km/h (9.9 mi/h) crosswind were to combine with a perpendicular 1600 km/h (994.19 mi/h) incoming airflow (i.e., 100 times as fast), it would only alter the incoming airflow direction by 0.573°. In contrast, long-range bullets can regularly follow trajectories that arc almost 90°. Therefore, the most significant cause of reorientation is due to gravity changing the trajectories of bullets and thereby the air resistance they face.

6.d Optimal Spin Rates

While bullets can be dynamically stable with an Sg = 1 in theory, they are **far more likely to be stable with an Sg's of at least 1.5** at the muzzle. (See Image 96, Pg. 98.) This 1.5 number is somewhat arbitrary, and generally ensures that even if bullets exit the muzzle with a bit of misalignment, their dynamic stability can overcome it. (See Image 97, Pg. 98.)

Tractability

Image 81: This artillery shell (essentially a very large bullet) impacted the tree at a downward angle, exemplifying the **forward turn** that bullets make as they travel.

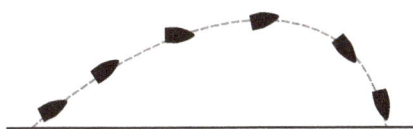

Tractable (Actual) Overstabilized (False)

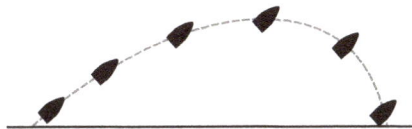

Image 82: Because a dynamically stable bullet always reorients into the incoming airflow, **it always reorients to its trajectory**. The myth of overstabilization comes from the concepts of gyroscopes, which resist reorientation. And bullets are gyroscopes. However, gyroscopes do not eliminate overturning force; they instead redirect it at a 90-degree angle into spin drift (See Spin Drift, Pg. 89.) or aerodynamic jump (See Aerodynamic (Muzzle) Jump, Pg. 86.). Spinning a bullet faster causes more of both, and makes bullets slower to turn forward (See Image 75, Pg. 75.), causing a negligibly higher drag coefficient that is accounted for with truing. (See Trajectory Validation for Ballistic Calculators (Truing), Pg. 68.)

Bullets are rarely intentionally spun faster than an Sg = 1.5 because additional spin cannot significantly get the tip pointed forward any faster and therefore cannot significantly increase the aerodynamics of the bullets. On the other hand, attempting a faster-than-necessary spin requires extra energy that could have been applied to the bullet's velocity instead. That said, bullets are regularly shot with Sg's > 1.5 when marksmen shoot one kind of ammunition out a rifle with a twist rate optimized for another kind of ammunition.

That said, **not much extra energy is required to spin a bullet faster**. As a practical example, a 5.56 mm (0.223 in) caliber rifle bullet experiences twice as many spins in a 1:152 mm (1:6 in) twist bore than in a 1:304 mm

(1:12 in) bore. When traveling 30 cm (12 in), the bullet's outer surface would travel 306.9 mm (12.08 in) with a 1:152 mm (1:6 in) twist, versus 305.4 mm (12.02 in) with a 1:152 mm (1:6 in) twist. That's a difference of about 0.5%. The radial midpoint of mass is about ⅓ of the radius below the surface; therefore, the average mass of the bullet travels less than 0.5% farther inside the bore when the twist rate is halved. That travel distance equates to less than 0.5% of the energy that is transferred from velocity to spin. For a bullet traveling at 800 m/s (2624 ft/s), doubling the twist rate causes a loss in muzzle velocity of about 2 m/s (6 ft/s). Empirical data roughly matches these numbers.

Excessive spin also carries two additional risks. First, a bullet's materials may experience such high stress due to the unnecessary rotational energy that the bullet catastrophically fails and bursts midair. Second, if mass were unevenly distributed in the bullet, then the center-of-mass would not be in the center of the bullet's shape. This discrepancy would cause a bullet to unstably rotate, or wobble, just like the buzz of a phone is caused by a motor spinning a lopsided weight. Increasing the spin would directly increase the wobble. That said, modern match-grade bullets are so well made and have such a small mass distribution error, that wobble is not a major concern.

Importantly, **over-stabilization is also not a concern**. The term "over-stabilization" refers to the idea that spinning a bullet too fast would cause it to maintain its the original orientation of its spin-axis, rather than reorienting the spin-axis to align with the incoming airflow and air resistance. In simpler terms, an over-stabilized bullet would spin so rapidly that its spin-axis would remain fixed in space. The farther this mythical overstabilized bullet were to travel along its curved trajectory, the more its tip would point away from the direction of travel, increasing air resistance and reducing efficiency.

But a bullet can never become fixed in space. This is because the reorientation comes from the loss of spin, and bullets that are spun faster have more spin to lose in the first place.

That said, **bullets that are spun faster are indeed slower to reorient** to the incoming airflow, causing increased drag. (See Image 75, Pg. 75.) This is because precession more effectively redirects the reorientation of the spin-axis 90 degrees, and not because reorientation is not occurring. (See Image 74, Pg. 75.)

But this **slower reorientation does not have a negative effect on accuracy nor precision** because the effect is predictable and repeatable, and so is accounted for by zeroing, ballistic calculators, and truing. In fact, while some experts claim that an Sg greater than 1.7 has is over-stabilized, other

Lift Force

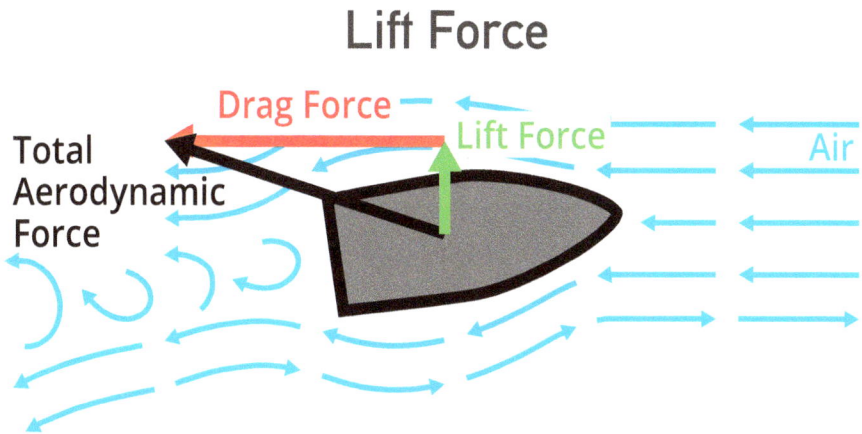

Drag Force

Lift Force

Air

Total
Aerodynamic
Force

Image 83: When an object **impacts air at an angle**, one side of the object contacts more resistance than the other side. This has two effects. First, the air directly pushes against that side more. Second, the air slows down more on the side it contacts more. Bernoulli's principle states that an decrease in the speed of a fluid occurs simultaneously with an increase in pressure, also pushing that side more. The combination of these two effects (direct and indirect pressure) cause an object to accelerate into the direction that the object is pointing.

experts regularly shoot bullets with an Sg of 2.3 or more with no decrease in precision.

The misconception about over-stabilization likely stems from analogizing bullets to gyroscopes, which are specifically designed to resist reorientation. A spinning top is a better analogy because they experience significant friction on their tips which reorients their spin-axis. Tops, like bullets, would have their materials fly apart due to the rotational energy of spinning long before the spinning caused destabilization.

6.e Vector Orienting to Bullet (Lift Force)

When an object moves through the air, it faces air resistance, a force that opposes its motion. Air resistance consists of two main components: drag and lift. Drag acts in the direction opposite to the object's movement, slowing it down. **Lift acts perpendicular to the airflow,** causing the object to change its path. (See Image 83, Pg. 83.) For example, aircraft use engines or jets to produce thrust, which counteracts drag. By overcoming drag, the aircraft allows lift to continuously act on it, pushing it upward to balance gravity and keep the plane in the air.

In simple terms, lift is generated when air exerts more force on one side of an object than the other. For bullets, lift occurs when they are not pointed exactly forward. The reasons why air pushes more on one side are highly complex. In addition to standard aerodynamic principles like Bernoulli's principle, supersonic fluid dynamics also play a role. However, this book does not explore those details, as the simple explanation is enough.

When experiencing lift, a bullet changes its trajectory away from the side experiencing more force until the air resistance is equalized between both sides of the bullet. In this way, **bullets are pushed into the direction they are pointing**. Put another way, the lift force pushes bullets into the path of least resistance in the air. Because a bullet maintains its narrowest orientation in only two ways (i.e., tip-forward or tail-forward), and a spin-stabilized bullet is already tip-forward, its direction-of-travel adjusts until it matches whichever direction the tip points.

The degree to which lift deflects a bullet depends on how effectively the bullet interacts with the air to create a difference in force on each of its sides. This force difference is influenced by several factors. Higher air density provides more air for interaction, increasing the force difference and enhancing lift. A greater speed differential between the bullet and the surrounding air causes each air molecule to impart more force, resulting in increased lift. Additionally, long, narrow bullets with a higher surface-area-to-volume ratio interact more with the air, further amplifying lift. In contrast, bullets in a vacuum, those traveling at the same speed as the surrounding air, and spherical objects do not experience lift.

6.f Precession-Induced 90-Degree Turning

As previously mentioned, precession is when the spin-axis of a spinning object rotates 90 degrees from the direction of an applied rotational-force (i.e., a torque). (See Image 74, Pg. 75.) In ballistics, two main factors can cause such rotating forces on a bullet's spin axis: crosswinds and gravity.

A crosswind is a wind that blows perpendicular to the bullet's path. By blowing into the side of a bullet, a crosswind changes the airflow around the bullet, affecting the thicker tail (more surface area) more than the thinner tip (less surface area). As a result, the bullet experiences a sideways rotation, altering its spin-axis. (See Image 84, Pg. 85.)

Gravity has a more subtle effect. As a bullet travels, gravity causes it to follow a curved trajectory. From the bullet's perspective, the air appears to

Spin-axis Rotation and Precession

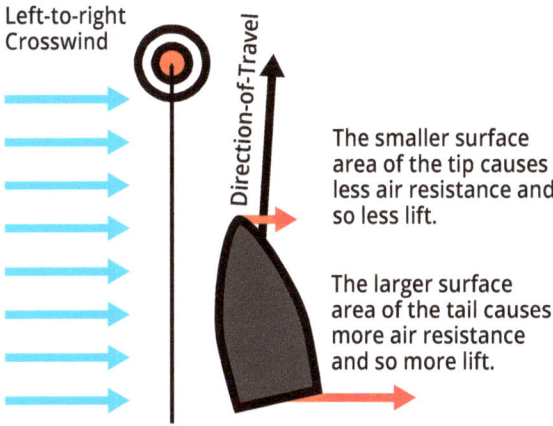

Image 84: The rear of a bullet has more surface area than the front and so the rear experiences **more force for the same pressure**, turning the bullet. Thereafter, precession redirects a portion of that force 90 degrees, pointing the bullet either more up or more down.

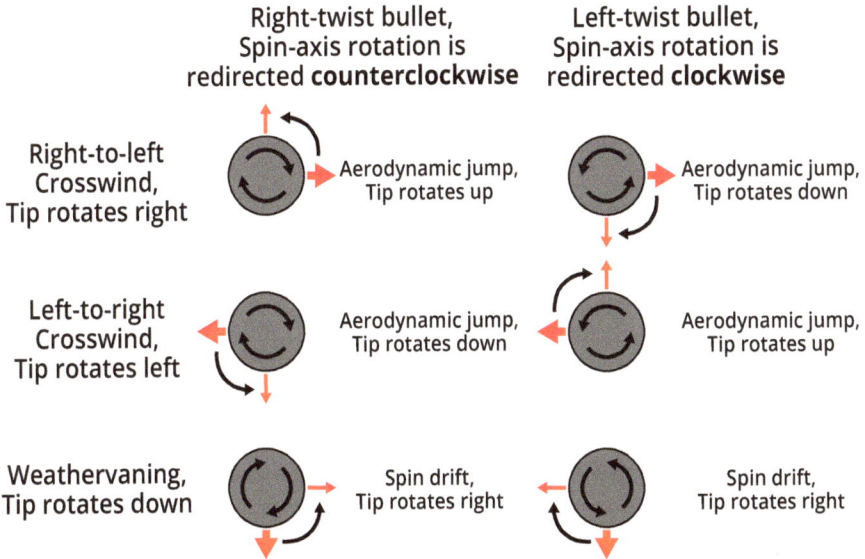

Image 85: This view is of the rear of a bullet. When a changing airflow applies a torque to a spinning bullet, that **torque is partially applied at a 90-degree angle**, resulting in either aerodynamic jump or spin drift. A right-twist bullet applies the torque 90 degrees counterclockwise (to the tip when viewed from the rear), while a left-twist bullet applies the torque 90 degrees clockwise.

accelerate upwards. This upward airflow increases the air pressure on the underside of the thicker tail compared to the tip, causing the spin axis to rotate forward.

When a crosswind or gravity applies a torque to a bullet, gyroscopic forces deflect part of that force to a 90-degree angle. Whether that 90 degrees is clockwise or counterclockwise depends on which direction the bullet is spinning. For a right-spinning bullet shot from a right-twist barrel, the deflection is to the right, and vice versa for a left-spinning bullet. (See Image 85, Pg. 85.)

The strength of the 90-degree torque is directly proportional to the strength of both the gyroscopic effect and the initial torque applied. The gyroscopic effect is determined by the bullet's spin rate (i.e., the gyroscopic stability factor). A faster-spinning bullet converts a larger percentage of applied torque 90 degrees and takes longer to realign with the incoming airflow. The aerodynamic force acting on the bullet depends on the wind strength or the steepness of the bullet's trajectory. The distance to the target can be relevant, but only insofar as steeper trajectories occur at farther distances.

So far, this explanation has described what makes a bullet point 90 degrees to the direction of an uneven application of pressure. However, it has yet to describe why a bullet that points in a particular direction would travel in that direction. **Bullets travel in the direction that they point because of lift.** (See Vector Orienting to Bullet (Lift Force), Pg. 83.)

An astute observer may notice that rotating the spin-axis 90 degrees would itself activate a second precession 90 degrees again (180 degrees in total). And this does in fact happen; however, precession only ever redirects a portion of an applied torque. Therefore, that second-order torque is small and masked by the stronger initial torque. In theory, there are infinite higher-order precessions of decreasing strength. However, because each successive iteration is only a fraction of the previous one, the net forces are only the initial reorientation of the spin-axis and the secondary 90-degree precession.

6.g Aerodynamic (Muzzle) Jump

Aerodynamic jump (a.k.a., muzzle jump) is the name for a precession-induced turn caused by a crosswind. It is called a "jump" because it occurs most strongly when the bullet leaves the muzzle and initially contacts and reorients to the wind.

Essentially, aerodynamic jump places a bullet on a higher or lower trajectory from the start (i.e., when it hits a wind). Therefore, a bullet experiences double the jump for double the crosswind. For example, a bullet

Aerodynamic Jump for a Right Twist

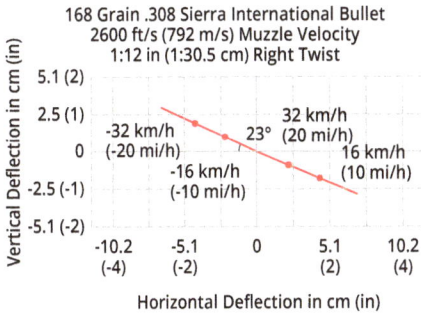

Image 86: At a given distance, aerodynamic jump occurs on a line. **A faster crosswind pushes and rotates a bullet more**; thereby precession redirects more of that rotation 90 degrees, allowing lift to push the bullet up or down more.

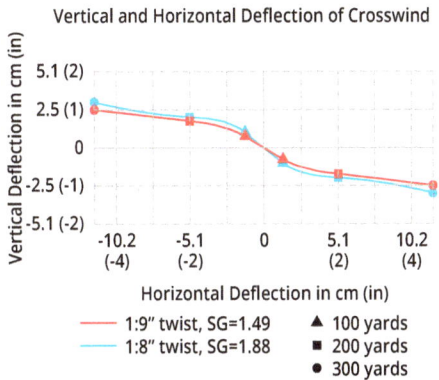

Image 87: **The slope of that line is proportional to the static stability**. Bullets with more spin, and thereby a higher gyroscopic (static) stability, redirect a larger portion of the crosswind's rotation 90 degrees, and so create a steeper slope.

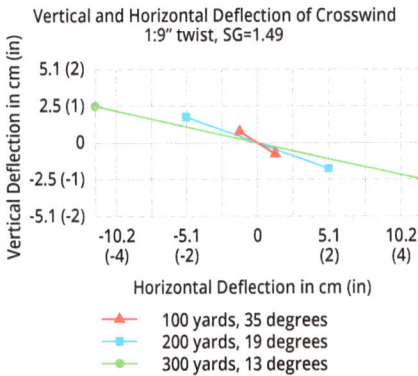

Image 88: As a bullet travels farther and slows down, the crosswind becomes a greater proportion of the bullet's total air resistance, shifting the vector of incoming airflow. The **largest jump occurs as the bullet exits the muzzle**, when the crosswind first hits and spins the bullet.

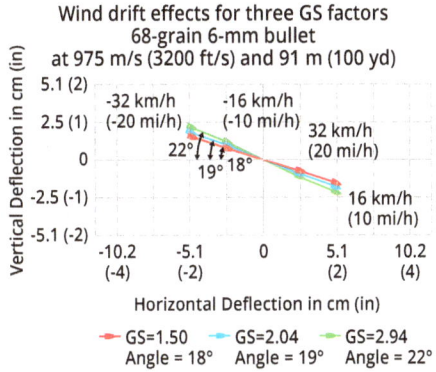

Image 89: This diagram combines the previous two. The largest jump occurs when the bullet exits the muzzle and meets the crosswind. Thereafter the jump is proportional to the velocity of the bullet. And **a shorter twist causes greater static stability, which in turn causes more jump**.

Aerodynamic Jump 2

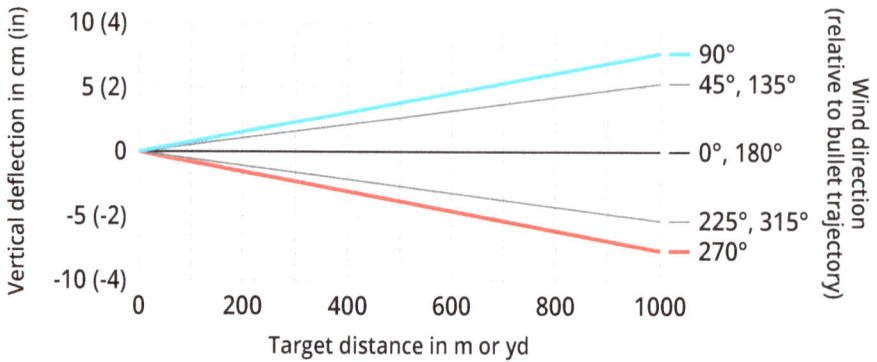

Image 90: This graph is generalized; the effect of aerodynamic jump is not truly a straight line. The purpose of this graph is rather to show that the effect of aerodynamic jump at even 1000 m or yd is only plus or minus a few cm or in.

that experiences 0.1 mils of jump with a 10 km/h or mi/h crosswind would experience 0.2 mils of jump in a 20 km/h or mi/h crosswind. Therefore, **aerodynamic jump for a weapon system can generally be defined at a mil deviation per 1 km/h or mi/h of effective crosswind**. Accounting for an aerodynamic jump can be especially important when truing a rifle at a far target in a strong crosswind.

For a right-spinning bullet, a right-to-left crosswind rotates a bullet down, and a left-to-right crosswind rotates a bullet up. (See Image 85, Pg. 85.) Therefore, a right-spinning bullet shot in different winds would form a straight line of bullet holes in a target from the top left through the center to the bottom right. (See Image 86, Pg. 87.) A mirrored line would appear for a left-spinning bullet or a vertical wind.

The angle of the aerodynamic-jump line depends on how fast the bullet is spinning (i.e., the percent of rotation translated 90 degrees), and how much lift the bullet experiences. (See Image 87, Pg. 87.) For example, shooting at 100 m or yd may create: an 18-degree line from a 35-cm (14-in) twist; a 19-degree line from a 30-cm (12-in) twist; or a 22-degree line from a 25-cm (10-in) twist. (See Image 87, Pg. 87.) This characteristic line is very familiar to precision marksmen at 100 to 200 m or yd when all other factors are accounted for.

The angle of the aerodynamic-jump line also depends on the distance to the target. (See Image 88, Pg. 87.) This is because the angle is determined by the ratio of the crosswind speed to the bullet's velocity, which reflects the direction of the incoming airflow. As bullets travel, they slow down. Consequently, a constant crosswind becomes a larger proportion of the incoming airflow, causing the bullet to rotate more. Increased torque results in additional aerodynamic jump. In other words, halving the bullet's velocity has the same effect as doubling the crosswind speed. Unfortunately, the relationship between the aerodynamic-jump line angle and distance to the target is not simple because bullet velocity does not decrease linearly.

This book uses the term "aerodynamic jump," but other sources might use different names for the same phenomenon. Additionally, some concepts like lateral throwoff are sometimes included under the umbrella of "aerodynamic jump," which can be confusing. Lateral throwoff is a separate phenomenon because it can occur in any direction, making its effect random rather than linear. Moreover, any non-random (predictable) lateral throwoff is virtually undetectable since it is already accounted for when zeroing the rifle.

In contrast, aerodynamic jump is significant because it can be consistently predicted using historical data and crosswind speed. Unlike lateral throwoff, which is random and varies in direction, aerodynamic jump follows a predictable pattern based on specific conditions, making it a more reliable factor to consider. (See Image 90, Pg. 88.)

6.h Spin Drift

Gravity causes bullets to experience a perpetually increasing updraft as gravity accelerates bullets to the ground. In other words, from the marksman's perspective, the air appears stable while the bullet's arc is curved; however, from the bullet's perspective, it remains stable itself while the incoming airflow takes on an increasingly upward velocity. The bullet always points slightly above the vector of incoming airflow as gravity constantly accelerates a bullet downwards, causing the direction of incoming airflow to constantly rotate. The result is a constant tiny overturning torque that tries to rotate the bullet's nose upward. For this reason, the act of the updraft caused by gravity pushing against the bullet's tip is sometimes referred to as the "overturning force."

Gyroscopic precession translates that torque on the spin-axis 90 degrees into a right or left rotation. (See Image 85, Pg. 85.) The resulting angle between the direction-of-movement and the spin-axis is called the "yaw-of-repose" or "angle-of-yaw." (See Image 74, Pg. 75.) A right-spinning and a left-spinning bullet would respectively have a right-facing and a left-facing angle-of-yaw.

Spin Drift

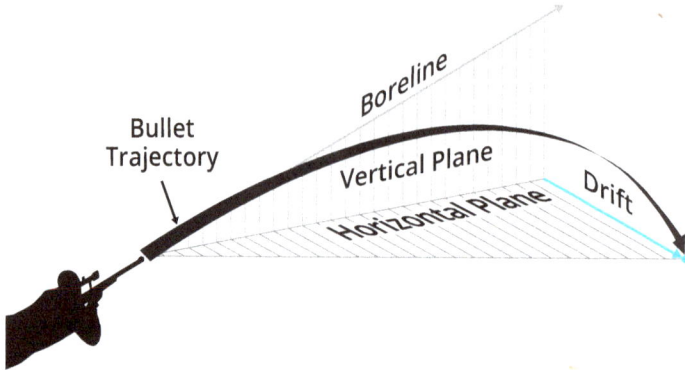

Image 91: Spin drift causes bullets to spin to the right for a right-twisted bullet and to the left for a left-twisted bullet. The drift is the induced **horizontal offset**.

Predicted spin drift for 5 bullets according to 3 programs at 3 distances

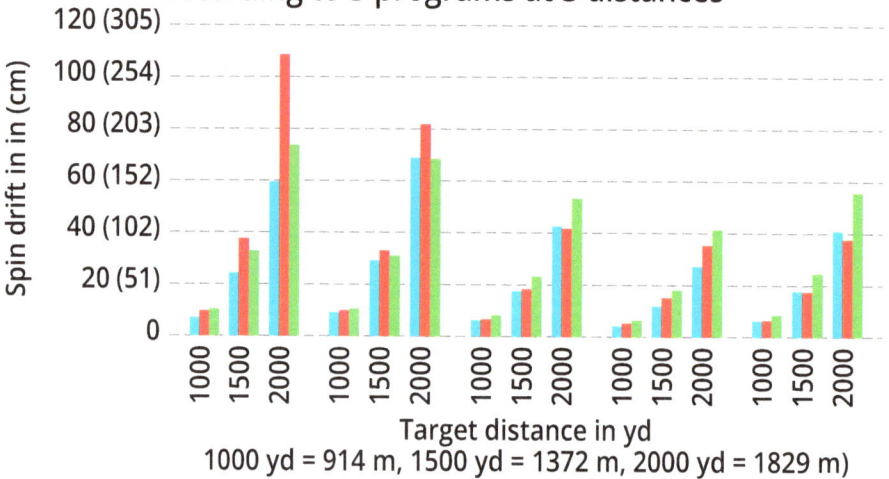

Target distance in yd
1000 yd = 914 m, 1500 yd = 1372 m, 2000 yd = 1829 m)

Bullets: 1) Sierra 168 Intl. · 2) 175 M118LR · 3) Sierra 220 SMK
4) Berger 215 Hybrid · 5) Hornady .338 285 BTHP

Prediction software: ■ B&R   ■ Hornady 4DOF   ■ Litz

Image 92: The amount of spin drift predicted by a ballistic calculator **depends on the formulas programmed** into that calculator. This graph shows predictions from three different softwares for five different bullets at three different distances. Clearly, the programs agree for the most part. At even 1000 m or yd, spin drift is still a minor effect, so the absolute differences between the programs only amount to a few cm or in. However as distance increases, the differences between the predictions become more pronounced and significant.

In sum, **spin drift uses the same principles as aerodynamic jump**, but with the shift in airflow being caused by falling down as opposed to by a crosswind. And just as with aerodynamic jump, once a bullet points in a new direction, lift causes it to go in that direction. (See Vector Orienting to Bullet (Lift Force), Pg. 83.) (See Image 90, Pg. 88.)

The amount of spin drift increases exponentially with distance. This exponential relationship occurs because spin drift is caused by gravity, which continuously accelerates bullets downward. As a result, the bullet's spin axis turns more and more over time, leading to an exponentially greater degree of spin drift the farther the bullet travels.

This complicated exponential relationship makes the calculation of spin drift mathematically complicated, and **many ballistic calculators are notoriously bad at calculating spin drift**. (See Image 91, Pg. 90.) In fact, many marksmen continue to turn off the spin drift calculation in their software because of that historical inaccuracy. Instead, these marksmen estimate that: spin drift is nominal below 500 m or yd; it equals between 1% and 2% of the elevation hold for slow and fast spinning bullets respectively at 1000 m or yd; and it increases exponentially thereafter. However, time has led to many technological advances and modern ballistic calculators are better at estimating spin drift than humans are.

The calculation of spin drift is made even more complicated by the fact that spin drift is influenced by many factors specific to each rifle system. For example, two identical cartridges fired from different rifles with different twist rates would experience different spin drift. And two bullets with different shapes fired from the same weapon system can also experience wildly different spin drift. At 1000 m or yd, a bullet can experience anywhere between 3 cm to 34 cm (1 in to 14 in) of spin drift, with most bullets experiencing between 20 cm and 30 cm (8 in and 12 in).

Another confounding factor is rifle cant. If a rifle were tilted to the right, a fired bullet would drift rightward. This cant-induced drift can be confused with spin drift because both result in significant sideways movement of a bullet at long ranges. However, the two types of drift can be distinguished. While both cant-induced drift and spin drift increase with respect to gravity, only spin drift also increases with respect to a change in a bullet's velocity. That means that only cant-induced drift appears at short distances. Therefore, when shooting at long ranges, marksmen must first perform a vertical target test (as explained in *Long Range Shooting: an Illustrated Manual*) at short range to ensure their scope is not canted.

The method for empirically measuring spin drift is rather ingenious. It involves using two identical rifles, with the only difference being that one has a left twist and the other a right twist. Both rifles fire the same ammunition simultaneously at a target. Since the twist direction is the only variable, any differences in where the bullets hit the target are solely due to spin drift. After firing many shots, the average distance between each pair of bullet impacts is calculated. This average distance is then divided by two to determine the spin drift for that specific weapon system under the given conditions. This approach effectively isolates and measures the effect of spin drift by accounting for all other variables.

7. More Air Effects

As a bullet travels through the air, several atmospheric factors can influence its trajectory. Crosswinds, which blow perpendicular to a bullet's trajectory, have an immediate and strong effect. Other factors, such as headwinds, tailwinds (winds that blow parallel to a bullet's path), and the transition from supersonic to subsonic speeds (i.e., the transonic range), take longer to affect a bullet. However, the effects of these slower-acting influences grow exponentially with distance, making them increasingly important to consider at very long ranges.

7.a Headwinds and Tailwinds (Parallel Winds)

Both headwinds and tailwinds blow parallel to a bullet's path. A headwind blows against the bullet's direction of travel, slowing it down. A tailwind blows from behind, pushing it forward. However, because bullets fly faster than any wind can blow, a bullet never truly experiences a "tailwind" from its own perspective. Instead, what feels like a tailwind to the marksman is simply a slower headwind to the bullet.

The faster that a parallel wind blows against the front of a bullet, the **more air resistance that the bullet experiences**. Therefore, stronger headwinds cause more air resistance. Conversely, tailwinds reduce air resistance, allowing a bullet to fly farther (given the same amount of initial energy). Notably, parallel winds only affect forward motion; a bullet falls to the Earth at roughly the same speed regardless of any headwind or tailwind.

Headwinds and tailwinds have a relatively small impact on a bullet's trajectory because these winds are minor compared to the bullet's forward

Headwinds and Tailwinds

Vertical change in point-of-impact (i.e., bullet-drop) versus distance a bullet travels due to a 25 k/h (10 m/h) wind incoming from 8 directions for a right-twist bullet

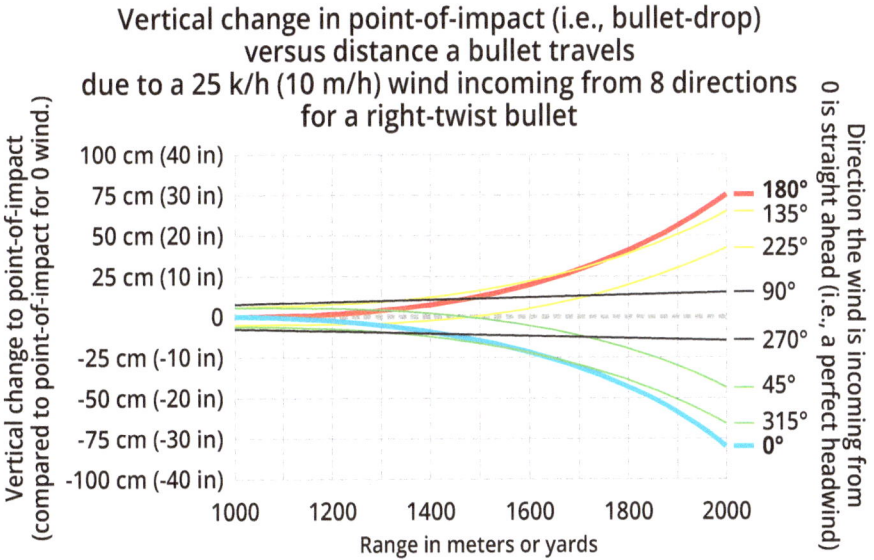

Image 93: Headwinds (blue, bottom, 0°) and tailwinds (red, top, 180°) do not affect the point-of-impact of a bullet until well after 1,000 m or yd. This effect can be seen in the diagram, where the red and blue lines do not diverge from the center until about 1100 m or yd for this particular ammunition. However, **the effect is exponential**, so it becomes vital to account for their effect thereafter. Crosswinds (0° and 270°) have a vertical effect due to aerodynamic jump (See Aerodynamic (Muzzle) Jump, Pg. 86.). The diagonal winds are a combination of the headwind-tailwind effect and the aerodynamic jump effect.

velocity. For instance, if a bullet travels at 1000 m/s against a headwind of 5 m/h, its combined airflow becomes 1005 m/s, meaning the wind is only 0.5% of the total. Even for a slower bullet traveling at 500 m/s, the same 5 m/s headwind still makes up just 1% of its airflow, and therefore just 1% of its total air resistance. Increased air resistance causes a bullet to drop faster than gravity alone would. Hence, headwinds and tailwinds only become significant when that 1-2% change in bullet-drop due to air resistance alone becomes large enough to matter.

Another way to frame the small effect of a parallel wind versus a crosswind is that **bullets are specifically designed to be aerodynamic** into an incoming airflow. That is, the shapes of bullets are designed to negate parallel winds.

To find a headwind or tailwind, a marksman must first measure the actual wind. Most of the time, winds are not perfectly parallel nor perfectly perpendicular; they instead approach from a diagonal angle with a crosswind component and a headwind-tailwind component. The crosswind component is frequently referred to as the wind value. In the same way, the parallel wind component can be labeled as either the headwind value or the tailwind value. Whereas the crosswind value is the cosine of the angle of the actual wind from the 12 o'clock direction, the parallel wind is instead the sine of that angle. Notably, because wind values are determined trigonometrically, the crosswind value and the parallel wind value sum to a number greater than one.

Unlike crosswinds, parallel winds are never calculated manually. Marksmen simply enter the wind's direction and velocity into a ballistic calculator, which automatically factors in any headwind or tailwind. Although a marksman could theoretically calculate parallel wind effects by hand, doing so for shots beyond 1 km or 1 mi would take so long that the wind would likely have shifted anyway.

7.b Transonic Range (Supersonic to Subsonic Shift)

Sound travels through air at a specific speed, commonly referred to as "Mach 1." The number after "Mach" indicates how many times the speed of sound an object is traveling, so Mach 2 is twice the speed of sound, and Mach 0.5 is half. Exactly what Mach 1 translates to in km/h or mi/h depends on environmental conditions. For example, sound travels faster in warmer air because air molecules move and transmit energy more quickly. It also moves faster at sea level, where air pressure is higher, since the molecules are closer together and interact more readily than at higher altitudes. Mach 1 at sea level and at 20°C (68°F) is **approximately 343 m/s (1125 ft/s)**, or about 1235 k/h (767 mi/h).

An object traveling slower than Mach 1 is called "subsonic," and an object traveling at or faster than Mach 1 is called "supersonic." The interactions of a subsonic object and a supersonic object with air are fundamentally different. This is because supersonic objects experience a strong form of drag called "wave drag." To understand wave drag, first, the fundamental cause of sound must be explained.

All sounds are made by the vibrations of molecules. For example, when a person hits a drum or a cymbal, the object deforms. Specifically, when an object is struck and depresses, a pocket of low air pressure is created, and

Transonic Range

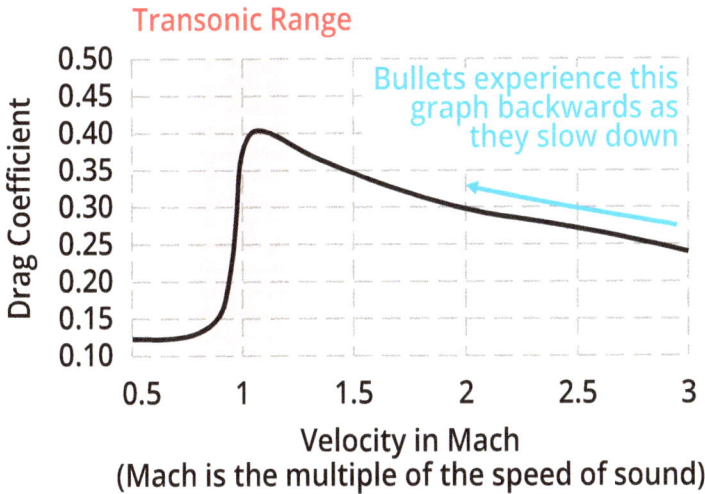

Image 94: This graph shows an example **relationship between the velocity of a bullet and its drag coefficient**. A higher drag coefficient means that an object experiences more resistance for a given airflow (e.g., a cube), and a lower coefficient means less resistance (e.g., a plane). The coefficient depends not only on an object's shape but also on its velocity. In both the subsonic and the supersonic range, the relationship shown here is relatively linear. However around Mach = 1, there is a sudden drop. This S-shaped section is called the "transonic range" and spans from Mach = 0.8 to Mach = 1.2 (because the range is defined by the complexity of the curve, the range is approximate). **The transonic range can be substantially different for different bullets** because of their different shapes, and so generalizing one curve to all bullets causes inaccurate results.

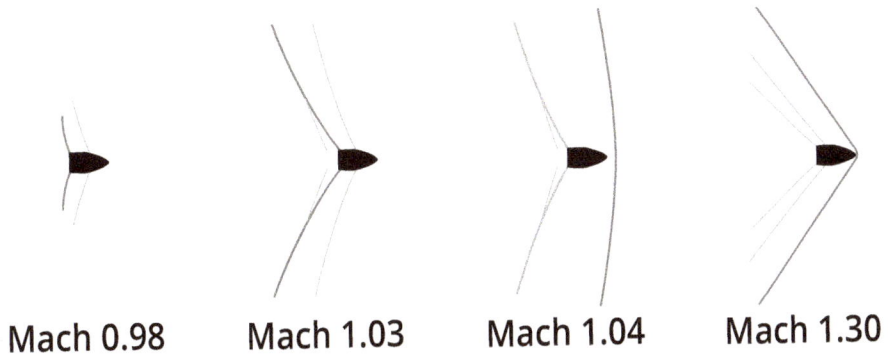

Mach 0.98   Mach 1.03   Mach 1.04   Mach 1.30

Image 95: The darker line represents **the pressure wave of supersonic air**. Due to complicated fluid dynamics, air surrounding an object moves at different speeds in different locations around the object. That means that in the transonic range, some air is supersonic while other air is subsonic.

when the object rebounds, a pocket of high air pressure is created. When objects depress and rebound in quick succession, that is called "vibration." When a bullet travels through the air, it creates an area of high pressure in front of it and a pocket of low pressure behind it.

These pockets or areas of high and low pressure travel outwards, away from their origin, and have the secondary effect of vibrating whatever they come into contact with. For example, when these pressure pockets vibrate an eardrum, a person hears those vibrations as "sound." Therefore, the speed of sound is the speed at which compressions and decompressions can travel through air (or any other material).

An object is considered supersonic when it travels faster than the speed at which pressure waves (or pockets of air pressure) can move outward. At these high speeds, **the pressure waves build up into shock waves** rather than dispersing. Because the pressure cannot spread out as it does at lower, subsonic speeds, it becomes more concentrated, increasing the drag on the object. This extra drag is known as "wave drag." It is clearly observed when a cone-shaped shock wave forms around a supersonic object. The sudden release of this shock wave creates a "sonic boom," which signifies a sharp jump in drag.

Although all long-range ammunition starts out traveling faster than the speed of sound, bullets can sometimes slow down to subsonic speeds before reaching their target. The transition from supersonic to subsonic does not happen abruptly right at Mach 1. Instead, it occurs gradually because **the air surrounding the bullet moves at different speeds and densities at various locations on the bullet**. For example, the air striking the bullet's nose might slow to subsonic speeds, while air flowing around the sides can still accelerate to supersonic speeds. This mixed-speed zone is called the "transonic range," and it generally spans bullet velocities between Mach 0.8 and Mach 1.2. (See Image 94, Pg. 95.)

The transonic range is often associated with a decrease in the accuracy and stability of bullets; however, this perception is only partly true. Air does not interact with bullets in a novel way that it does not in the subsonic or supersonic ranges. Instead, as the bullet moves from supersonic to subsonic speeds, the airflow around it becomes more complex, causing rapid shifts in the drag coefficient and other aerodynamic properties. Also, external forces have more time to act on a bullet per unit distance traveled because transonic bullets travel slower than supersonic bullets. Therefore, while the trajectories of transonic bullets are more difficult to predict, their trajectories are not inherently more random or chaotic. In physics, "chaos" means a small change in initial conditions can lead to a large change in outcome, and so the opposite of chaos is repeatability. The transonic range demonstrates this repeatability: **when two identical bullets are fired under the same conditions, they generally follow the same transonic trajectory**.

In fact, bullets experience a slower loss of velocity once they transition into the subsonic range, where they are traveling below the speed of sound. This occurs because the rate at which a bullet decelerates is tied directly to its drag, and drag itself is largely a function of velocity. When a bullet is moving at higher, supersonic speeds, it encounters a greater amount of drag; as a result, it slows down more rapidly. However, once the bullet's velocity drops below the speed of sound, the drag forces become less intense, causing the bullet to decelerate at a slower rate. Therefore, bullets in the subsonic range maintain their speeds more effectively over a given distance compared to bullets traveling faster. For example, a bullet may lose more absolute velocity in the first 100 m or yd of supersonic flight than in 500 m or yd of subsonic flight.

The reputation of transonic flight as being less accurate likely stems from the inability to generalize the transonic curve of one type of bullet to another kind of bullet. More specifically, when drag (Y-axis) is plotted against bullet velocity (X-axis), the resulting "drag curve" (See Image 94, Pg. 95.) shows how much air resistance a bullet experiences at any given speed. Even small differences in bullet geometry can greatly alter this curve, particularly in the chaotic transonic range. Therefore, any standardized drag model, such as G7 and G1 are relatively inaccurate and unhelpful in the transonic range.

A solution to this lack of generalizability is **custom drag models**, where each kind of bullet starts with a drag curve that is based on that bullet's particular geometry. However, even custom drag models are not perfect. They do account for how varying climate conditions can also affect a drag curve. For example, because the speed of sound increases when temperature increases, bullets enter and exit the transonic range sooner in their trajectories (i.e., at a higher velocity) in hotter climates than in colder ones. Therefore, truing is still necessary. (See Determining Bullet Trajectory, Pg. 60.)

A marksman can improve their performance in the transonic range by decreasing the effect of air resistance in the first place. Bullets with a high ballistic coefficient (i.e., low drag) are long and narrow, allowing them to slice through the air more efficiently than blunt bullets and have a smaller sonic boom. Further, because air is less compressed for a narrow bullet and it is less disruptive than a blunt bullet, a narrow bullet design has a

Static versus Dynamic Stability Factors

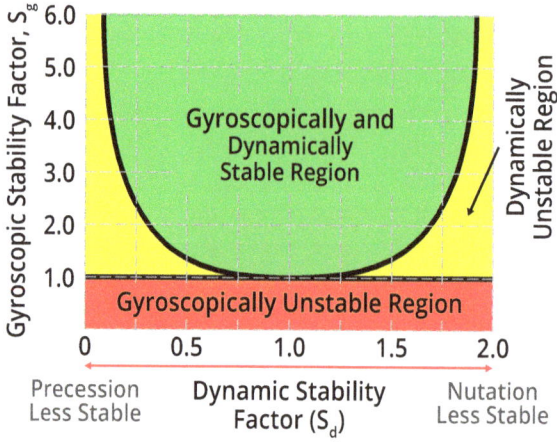

Image 96: **Gyroscopic (static) stability** is the ability of a bullet to have its tip point forward on average. By definition, a tip points forward on average over time when the static stability factor is above 1. Bullets that spin faster at a given velocity have higher static stability than those that spin slower.

Dynamic stability is the ability of the tip of a bullet to align to the incoming airflow. In contrast, a dynamically unstable (but still statically stable) bullet's tip rotates in larger and larger circles around the axis of travel as it moves forward. The dynamic stability factor determines whether a bullet is dynamically stable or not. A change in bullet spin does not have a large effect on dynamic stability.

This graph shows **three regions**. In the green, middle region, bullets are both statically and dynamically stable. Therefore, they travel with a constant, minimal drag. In the red, bottom region, bullets are not statically stable and so they tumble. In the yellow, side region, bullets are not tumbling, but their orientation is constantly changing. Drag depends on orientation. (Bullets are only aerodynamic if they face directly forward.) Therefore, **a changing orientation complexly increases drag and eventually leads to tumbling.**

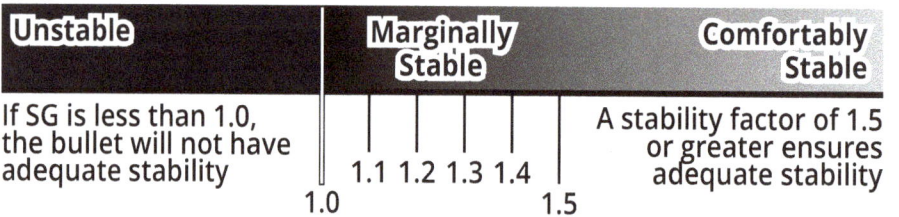

Image 97: The above relationship between stability factors is why most marksmen use twist rates that generate at least a static stability factor of at least 1.5. A marginally stable bullet with Sg < 1.5 can **easily become dynamically unstable**.

Falling Velocity's Effect on Stabilities

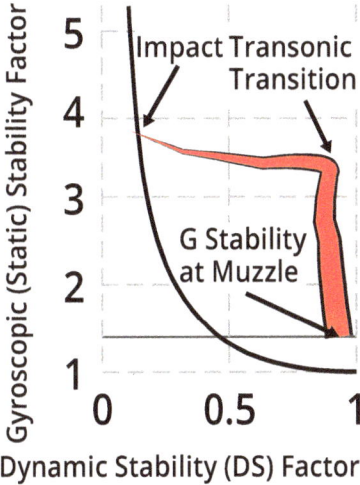

Image 98: The tapering line is the bullet's falling velocity. Going transonic causes **a sharp decrease** in the DS factor. Once the velocity line crosses the frontier line, the bullet loses dynamic stability.

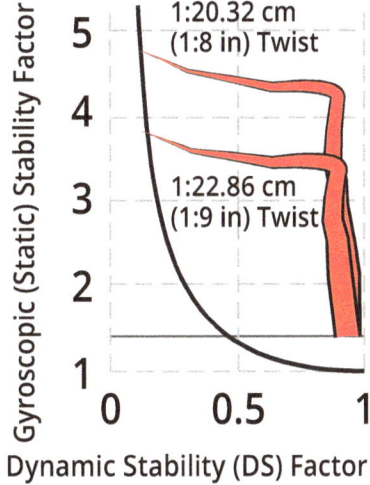

Image 99: The frontier line is a parabola, so increasing static stability does allow the DS factor to decrease more before dynamic stability is lost. But increasing static stability has **exponentially decreasing returns**.

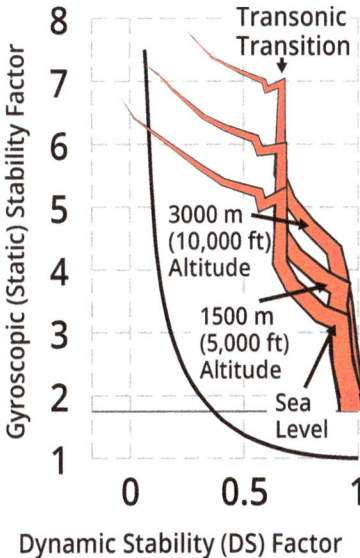

Image 100: Decreasing air density lowers the transonic threshold speed.

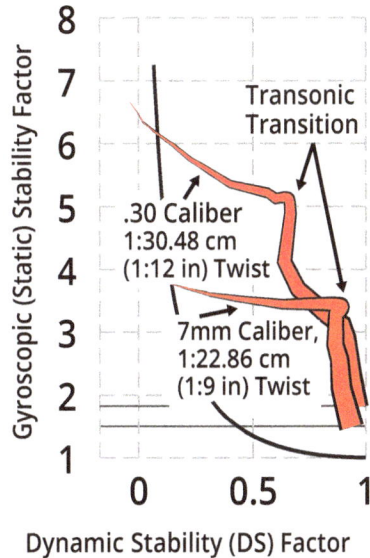

Image 101: Different bullets can have **vastly different stability curves**.

narrower transonic range. Therefore, narrower bullets retain their velocity for a longer time.

However, once narrower bullets do become unstable, they tumble far sooner. Longer bullets catch more air on their sides (per unit weight) if their tips deviate from pointing forward. In more technical terms, narrower bullets are more susceptible to dynamic instability because the center-of-pressure moves forward through the transonic range, increasing the overturning torque applied to the bullet. Dynamic instability means that a bullet's tip still points forward on average; however at any instant, the tip is slowly diverging from the bullet's direction of travel (i.e., angle-of-yaw increases). Dynamically unstable bullets eventually tumble, which massively increases air resistance and decreases predictability. Therefore, **long bullets better retain velocity but short bullets better retain stability**.

For example, when comparing a short 250-grain match bullet with a longer, 300-grain bullet, the 300-grain bullet performs better in the supersonic range. Moreover, the 300-grain bullet experiences a slower loss of velocity and slows to the transonic range at a longer distance. However, the shorter 250-grain bullet has a higher stability and endures the transonic range better than the longer, heavier bullet. Another example is that a boat tail of greater than 10 degrees makes a bullet more aerodynamic, but also narrower and less stable in the transonic range. This tradeoff between better supersonic performance and better transonic performance is a common consideration when choosing the optimal bullet for long-range shooting.

Another method to counter the extreme drag of the transonic range is to simply spin a bullet faster. Faster gyroscopic stabilization can overcome the lower dynamic stability inherent to longer bullets traveling in the transonic region. However, spinning bullets has diminishing returns. Increasing the twist rate by 10% may potentially sustain stability for perhaps an extra 5% of distance or less. (See Image 99, Pg. 99.) In fact, increasing the rifle twist rate for extremely lengthy bullets may yield no results at all.

Again, the exact relationship between velocity and stability varies greatly depending on the bullet because the effects of the transonic range are so sensitive to minute change. (See Image 101, Pg. 99.) If transonic information for a specific rifle system cannot be obtained from a manufacturer nor from a ballistic calculator, **trial-and-error** usually yields the best results, even if it is tedious to do.

8. Acceleration on a Rotating Body

The Coriolis effect is a pseudo-force that causes bullets to drift to the right in the Northern Hemisphere and to the left in the Southern Hemisphere. This effect becomes more pronounced when a marksman is farther from the equator and closer to either pole.

It is considered a pseudo-force rather than a true force because the apparent drift results from two interacting factors: first, the Earth's surface and everything on it are constantly rotating; and second, a bullet takes time to reach its target. Due to these factors, the target that a bullet was fired towards moves from its initial position by the time the bullet arrives. This movement makes the bullet appear deflected from where it was originally aimed at. In reality, the bullet is not being forced to drift; instead, the target moves out of position during the bullet's flight.

Although a simplified mathematical explanation is provided below, marksmen do not perform complex Coriolis calculations on the shooting range. Instead, they simply plug geographic data into a ballistic calculator to get the correct outputs. Calculators are necessary because the Coriolis effect is so small and complex. For example, depending on the rifle and ammunition, the Coriolis effect only causes about a 10-cm (4-in) deflection at a distance of 1000 m or yd. (That said, as the distance to the target increases, the effect grows exponentially since it scales linearly with both the distance to the target and the bullet's flight time.)

8.a Latitudinal (North-South) Coriolis Effect

Latitude is a geographic coordinate that indicates a point's North-South position on Earth's surface. When a bullet is fired toward either the North or South Pole, it experiences an eastward deflection. Conversely, when a bullet is fired toward the equator, it deflects to the West. This phenomenon means that bullets in the Northern Hemisphere are deflected to the right, while those in the Southern Hemisphere are deflected to the left.

The reason why bullets deflect when fired across latitudes is because the bullet becomes physically closer to Earth's spin-axis. The distance from the spin-axis is important because every part of the Earth completes one rotation in the same amount of time: one day. To achieve this uniform rotation period, the equator must spin much faster than the poles. This is necessary because objects located farther from the spin axis must travel a larger circular path compared to those closer to the axis. Specifically, the rotational speed at the equator is about 464 m/s (1522 ft/s), whereas the rotational speed at the poles is nearly zero.

A bullet fired from one latitude to another latitude retains the rotational speed of its original latitude circle. In other words, due to the conservation of angular momentum, the bullet continues to rotate around the Earth at the same speed regardless of when the Earth's rotation becomes faster (closer to the equator) or slower (closer to the poles). (See Image 101, Pg. 99.)

This is why shooting towards the poles (i.e., into a smaller rotation circle) deflects to the East: because the bullet would rotate faster than the surface underneath it. Similarly, shooting towards the equator (i.e., into a larger rotation circle) deflects to the West because the bullet would rotate slower than the ground underneath it. (Moving East or West is by definition to have a different rotational speed compared to the Earth's surface.)

When imagining a change in latitude, it is important to remember that bullets are fired tangentially to the Earth's surface. (See Image 103, Pg. 105.) So for example, if a marksman in Portland, Oregon (located at ~45° latitude) were to fire a bullet North or South, it would travel at about a 45° angle towards the North Pole or the equator respectively. Therefore, the Coriolis effect is more powerful as marksmen get closer to the poles, and the surface of the Earth is more perpendicular to the Earth's spin-axis.

Coriolis drift is calculated with the formula:

▸ Coriolis Drift = $\omega \times \sin(\text{Latitude}) \times (D^2 \div V)$. (Technically, this is only valid when vertical motion is negligible, and calculators may use more complicated three-dimensional formulas, but this is sufficient for demonstrative purposes.)

▸ ω represents the Earth's rotation rate, which is 0.00007292 radians/second (rad/s). There are 2π (~6.28) radians in a circle. (A radian is a unit of angular measure defined by the ratio of an arc length to its radius. Since both the arc length and the radius are measured in the same unit, their units cancel each other out. As a result, a radian is considered a dimensionless quantity.)

▸ Latitude is measured in degrees from the equator, with positive values indicating positions in the Northern Hemisphere and negative values in the Southern Hemisphere. By applying the sine function to the latitude, the component of Earth's rotation that affects an object at a given latitude can be determined. This mathematical relationship allows for the calculation of how Earth's rotational speed influences objects differently depending on their North-South position on the planet.

Latitudinal Coriolis Effect

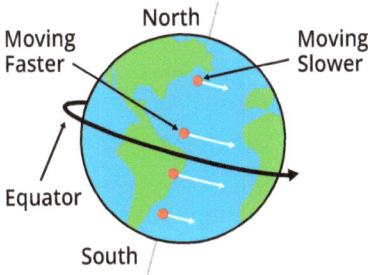

Objects closer to the equator move slower than objects closer to the equator

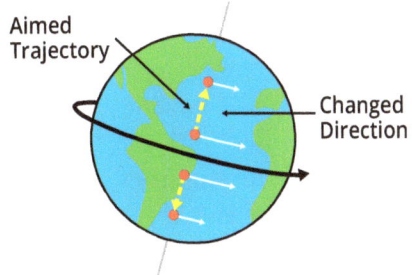

A Bullet is fired at a target at a slower-moving location.

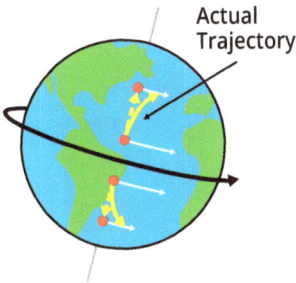

The object retains its initial, faster eastward velocity when it arrives at its final destination, farther from the equator.

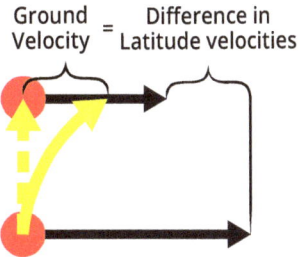

Its final ground speed is equal to difference between its intial rotational velocity minus its final rotational velocity.

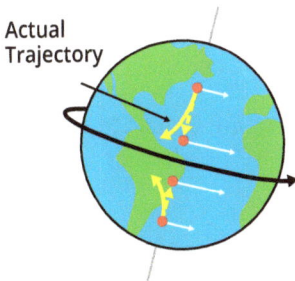

The object retains its initial, slower eastward velocity when it arrives at its final destination, closer to the equator.

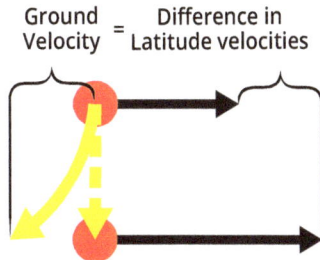

Its final ground speed is equal to difference between its intial rotational velocity minus its final rotational velocity.

Image 102: When a bullet is fired in the North-South direction, the Coriolis effect occurs because bullets retain the rotational velocity they have at their original location when they arrive at their new location at a different latitude.

▸ D represents the distance that the bullet travels. Both the distance and the bullet's velocity must use the same linear units (here, meters). Because the distance to the target is a squared variable, the effect quickly increases at longer distances.

▸ V is the bullet's average velocity from start to finish. (The Coriolis effect, which arises from the conservation of angular momentum, causes the bullet to deviate from its initial path. Importantly, the actual trajectory that the bullet follows does not impact the calculation of this deviation.)

▸ (A simplification of (D^2 / V) is (distance × flight time); however, using (D^2 / V) better illustrates why the Coriolis effect is only significant at longer ranges.)

Consider a bullet fired at Anchorage, Alaska (61.22° N) to a target 1000 m due North, at a speed of 1000 m/s.

▸ Total Coriolis drift = ω × sin(Latitude) × $(D^2 \div V)$

▸ 0.00007292 rad/s × sin (61.22°) × (1000 m)² ÷ 1000 m/s

▸ 0.00007292 rad/s × 0.876475 × 1000 m × 1 s

▸ 0.06391 m ≈ 6.4 cm.

The bullet would experience about 6.4 cm of Coriolis effect.

8.b Longitudinal (East-West) Coriolis Effect

A longitude is a geographic coordinate that specifies the East-West position of a point on Earth's surface. When a bullet is fired eastward, it deflects toward the equator. Conversely, when fired westward, it deflects toward the North Pole in the Northern Hemisphere or toward the South Pole in the Southern Hemisphere. In other words, bullets in the Northern Hemisphere are deflected to the right, while those in the Southern Hemisphere are deflected to the left.

In fact, the equation for longitudinal Coriolis effect and latitudinal Coriolis effect are identical:

▸ Coriolis Drift = ω × sin(Latitude) × $(D^2 \div V)$.

This can be reinforced with a thought experiment. At the poles, East-West shooting merges into North-South shooting because there is neither an East nor a West at the poles.

The East-West Coriolis effect arises for a more subtle reason than its North-South counterpart. When a bullet is fired at a target situated on the same latitude as its origin (i.e., traveling across longitudes), it does not remain on that latitude for the entire flight (unless it is at the equator). Because

Longitudinal Coriolis Effect 1

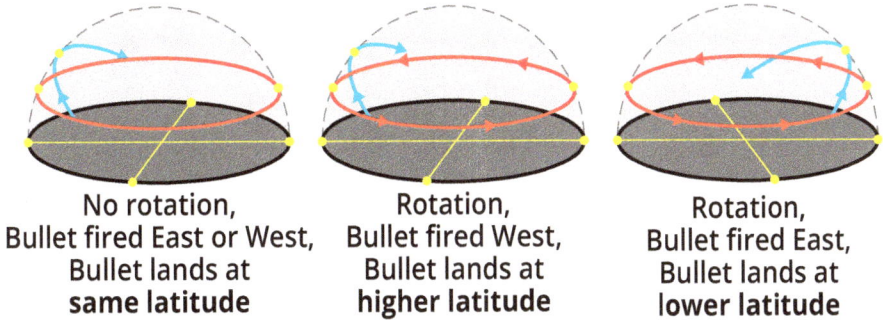

| No rotation, Bullet fired East or West, Bullet lands at **same latitude** | Rotation, Bullet fired West, Bullet lands at **higher latitude** | Rotation, Bullet fired East, Bullet lands at **lower latitude** |

Image 103: Bullet trajectories follow the curve of the Earth, and not lines of longitude. Therefore, for a marksman to hit a target on the same latitude as themself, their bullet must travel through latitudes closer to the poles before falling back down to the original latitude. **Firing a bullet westward reduces the centrifugal force applied to the bullet**, preventing it from fully returning to its original latitude. In contrast, firing a bullet eastward increases the amount of centrifugal force experienced, propelling the bullet closer to the equator. This diagram shows the same effect as the following diagram, but in a more three-dimensional way. (See Image 104, Pg. 106.)

---

bullets must counteract gravity to travel, they follow an arc that is tangential to Earth's surface at their location. As a result, in the Northern Hemisphere, a bullet's path shifts into latitudes closer to the North Pole before returning to the latitude of the target roughly halfway through its flight; in the Southern Hemisphere, it moves toward latitudes closer to the South Pole and then falls back to the target's latitude. (See Image 103, Pg. 105.)

When a bullet is fired eastward, the Earth's rotation adds to its angular momentum, effectively pushing it outward and increasing the centrifugal force it experiences. As a result, the bullet tends to overshoot its intended latitude, dropping through additional latitudes and ultimately ending up closer to the equator.

Similarly, **firing westward opposes and cancels out the rotation of the Earth**, causing the bullet to not fall as far as it needs to return to its original latitude. Therefore, it contacts the Earth earlier and remains at a higher latitude. (See Image 104, Pg. 106.)

These two examples assume that the stopping point, target, or destination is a location on the ground that is visible to the marksman. This assumption is important because if a westward-fired bullet is fired fast enough, it could enter an orbit and continue to a lower latitude than the original target is on (albeit at a lower velocity than an identically fired eastward-bound bullet).

Longitudinal Coriolis Effect 2

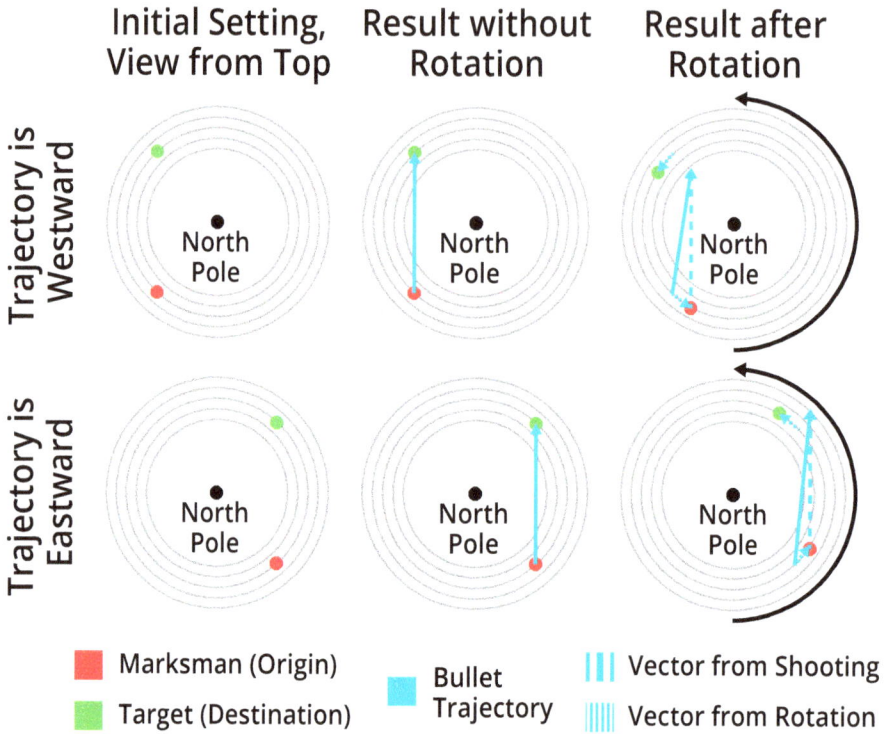

| Initial Setting, View from Top | Result without Rotation | Result after Rotation |

Image 104: Bullets are fired perpendicular to the ground. In contrast, lines of latitude are perpendicular to the Earth's spin-axis. Therefore, a fired bullet must traverse latitudes to go from the muzzle to the target. In the Northern Hemisphere, such a bullet briefly moves to a higher latitude before dropping back down.

On a rotating Earth, a bullet already has velocity before being fired. Whether the bullet is fired westward or eastward affects how its velocity combines with the Earth's rotation. A bullet that is fired westward is already moving away from the target's location, and a bullet fired eastward is already moving towards the target's location. This causes a westward bullet to have a slower velocity relative to the ground; therefore, it remains at a higher latitude instead of falling down to its original latitude. In contrast, an eastward-fired bullet has the bullet's velocity and the rotational velocity added together, so it overshoots its original latitude to end at a lower latitude than it was at originally.

Because the Earth rotates to the East, a bullet in the Northern Hemisphere always impacts to the right of the target, and a bullet in the Southern Hemisphere impacts to the left.

There are some other minor causes of the Coriolis effect. For example, the Earth is not a perfect sphere, and is actually 43 km (27 m) wider around the equator than around the poles. However, that is negligible for the Coriolis effect as it applies to bullets.

8.c Altitude (Up-Down) Eotvos Effect

Previously, the North-South and East-West Coriolis effects were discussed. They both describe how objects behave as they move on the Earth's surface. However, there is a third Coriolis effect, called the Eotvos effect, that describes how objects move in the third dimension: towards and away from the Earth's surface.

In sum, **the Eotvos effect causes a bullet fired to the East to land higher on a target than the initial point-of-aim, and a bullet fired to the West to land lower than the initial point-of-aim**. The Eotvos effect occurs in ballistics because the Earth rotates eastward during a bullet's flight time. (If a bullet were fired westward and instantly arrived, it would impact exactly on the point-of-aim since the target would have no time to advance into the bullet.)

As a bullet flies eastward, its target rotates away from the bullet; and as a bullet flies westward, the target rotates towards the bullet. It helps to imagine the bullet as flying in a straight line like a laser. Rotating into a straight line leads to a lower point-of-impact, whereas rotating away from the straight light raises the point-of-impact.

Another way to understand this phenomenon is by considering the bullet's perspective in terms of rotational speed. The bullet, when at rest, rotates in sync with Earth's rotation and thus experiences centrifugal force. Firing the bullet eastward increases its rotational speed and the associated centrifugal force, causing the bullet to gain additional centrifugal force. Conversely, firing it westward reduces its effective rotational speed and centrifugal force, allowing gravity to pull the bullet downward more forcefully. (See Image 105, Pg. 109.)

In contrast to the two other Coriolis effects, the Eotvos effect is strongest at the equator and is zero at the poles. However, much like the other Coriolis effects, the Eotvos effect is not very strong. There is a centrifugal force equal to only 0.34% of the force of gravity at the equator. Therefore, even eliminating the centrifugal force entirely doesn't have a large effect.

The formula for the Eotvos effect is:

- Eotvos effect =   ((*Distance-to-target*) × (*Time-of-flight*)) ×
                      (ω × cos(*Latitude*) × sin(*Azimuth*)).

Because time-of-flight can be difficult to find, the formula can be rearranged.

- (*Distance-to-target*) × (*Time-of-flight*) = ((V × 2 × (*Bullet-Drop*)) ÷ g)
- Eotvos effect =   ((V × 2 × (*Bullet-Drop*)) ÷ g) ×
                      (ω × cos(*Latitude*) × sin(*Azimuth*)).
- V is the bullet's average velocity.
- (*Bullet-Drop*) is the expected distance a bullet vertically drops at the target distance. This number can be found by firing North or South, where there is no Eotvos effect.
- g is the acceleration due to gravity (9.8 m/s^2).
- ω represents the Earth's rotation rate (0.00007292 radians/second).
- (*Latitude*) is degrees from the equator.
- (*Azimuth*) is the direction in which the bullet is fired, clockwise from true North.

So for example, if a bullet were to travel 1000 m in 1 s, at 0 degrees latitude (i.e., the equator), directly West (270 degrees), the formula with units omitted would be:

- Eotvos effect =   (1000 × 1) ×
                      (0.00007292 × cos(0) × sin(270))
                    ≈ -0.073 m ≈7 cm below expected.

Using the second formula, assuming a drop of 4.9 m (16 ft) and a velocity of 1000 m/s at 0 degrees latitude (i.e., the equator) and directly West (270 degrees), the formula with units omitted would be:

- Eotvos effect =   ((2 × 4.9 × 1000) ÷ 9.8) ×
                      (0.00007292 × cos(0) × sin(270))
                    ≈ -0.073 m ≈7 cm below expected.

For this example, a bullet would drop ~7 cm (2.8 in) more than would be expected without the Eotvos effect. Using the same numbers except shooting East would result in 7 fewer centimeters of drop, since sin(90) is positive whereas sin(270) is negative. Therefore for these numbers, there is a ~14 cm (5.5 in) differential in drop when shooting East compared to West.

Altitude Eotvos Effect

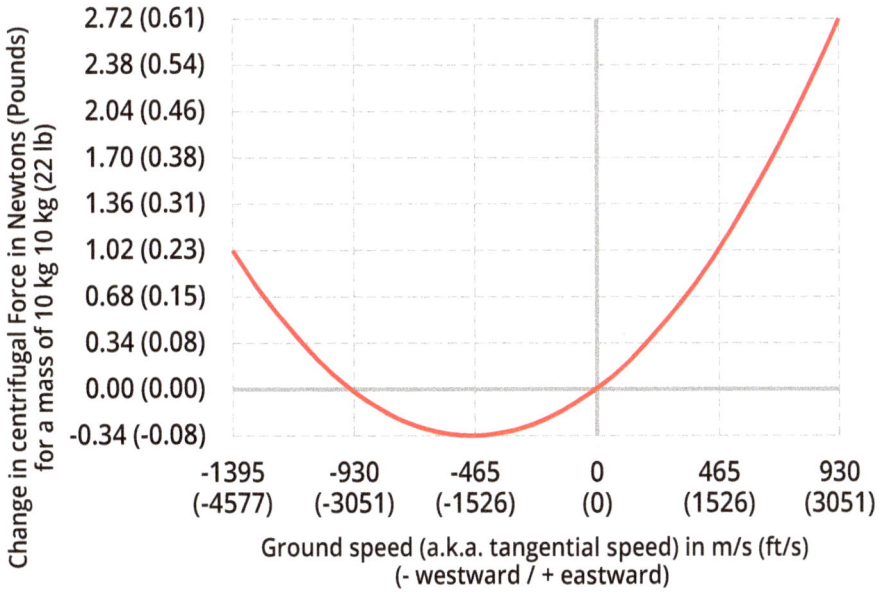

Image 105: The origin of this graph represents the baseline centrifugal force an object experiences on Earth's surface when at rest. Adding eastward or westward velocity to an object changes the amount of centrifugal force that object experiences.

The Eotvos effect is closely tied to the longitudinal Coriolis effect. As a bullet travels westward, the velocity of the shot **effectively cancels out** a portion of Earth's eastward rotation. At the equator, where Earth spins at 465 m/s (1526 ft/s), firing a bullet westward at that same speed would reduce the bullet's angular velocity to zero, thereby eliminating its centrifugal force. If a bullet is fired even faster in the westward direction, a counter-rotation begins, reintroducing a different centrifugal force. Because the Eotvos effect depends on Earth's rotational speed, the effect diminishes with increasing latitude, where local rotational speed is lower.

De Minimis Contents

De Minimis Effects

De minimis non curat praetor. ("A praetor is unconcerned with trifles.")
—A legal principle dating back to at least the 15th century

"De minimis" means too trivial or minor to merit consideration. Many ballistic effects exist that simply do not affect the trajectory of bullets enough that they are worth calculating with modern technology. These de-minimis effects can be ignored because humans cannot physically hold rifles still enough or with enough precision to adjust for these effects.

Even if a bullet were fired from a robotic setup, the theoretical marginal benefit would be extremely slight to the point that it would be an **inefficient use of time** to code a program or input the data. This time inefficiency exists in large part because many of these effects are accounted for by other processes. For example, almost all effects of local gravity are accounted for by simply re-zeroing and re-truing in the local area in which the marksman plans to shoot. In fact, these effects are so small that attempting to account for them could introduce additional errors if the program overcorrects.

Finally, even if a robot did exist that could accurately account for these effects, the result would not be to definitively change any bullet's trajectory from a miss to a hit. A bullet's trajectory at the distance these effects matter passes through kilometers or miles of randomized wind that would unpredictably alter the trajectory anyway. Therefore, accounting for these effects would only increase the probability of a hit (by a microscopic amount).

In sum, perfectly accounting for these effects would result in a bullet being ever so slightly more likely to be on target in the same way that buying a lottery ticket technically makes one more likely to win the lottery. That is, **it would be technically true but a waste of time.**

9. Local Gravity

Gravity is not the same everywhere on Earth. The acceleration due to Earth's gravity on its surface varies from 9.76 to 9.83 m/s² (32.02 to 32.25 ft/s²), which amounts to a range of about 1%. This difference exists for a few reasons. For example, gravity is about 0.3% stronger at the poles than at the equator due to the influence of centrifugal force. (The equator rotates at about 460 m/s (1509 ft/s), while the poles rotate at 0 m/s (0 ft/s.)) That is, as

111

the Earth rotates, it throws everything on its surface into space, counteracting a small percentage of gravity.

Gravity is also less at the equator because the Earth is not a perfect sphere. The same centrifugal force squishes the Earth out at the equator, and so objects at the equator are actually farther from the Earth's center-of-mass than objects at the poles. These mass effects can get very complicated because the ellipsoid shape of the Earth adds mass at the equator and removes mass from the poles. In brief, the shape of the Earth can cause objects at the poles to be 0.6% heavier at the equator.

Further, local terrain can affect gravity in a variety of ways. For example, gravity differs depending on the local altitude. For example, the difference in gravity from the top of Mount Everest to sea level is about 0.2%. Gravity also varies depending on the density of local material by about 0.1%. This fact was first proven by the Schiehallion experiment, an 18th-century experiment that was performed to determine the mean density of the Earth. The experiment measured how much the gravity of a nearby mountain could deflect a tool sideways. In sum, gravity is more intense near expansive areas of rock, such as mountain ranges. (See Image 106, Pg. 113.)

In order of the largest effect to smallest effect, with a spherical, homogeneous earth as a hypothetical base gravity, the variations from that base gravity and their scale would be:

- 10^{-1}% flattening, centrifugal acceleration
- 10^{-2}% mountains, valleys, ocean ridges, subduction
- 10^{-3}% density variations in crust and mantle
- 10^{-4}% salt domes, sediment basins, ores
- 10^{-5}% tides, atmospheric pressure
- 10^{-6}% temporal variations: oceans, hydrology
- 10^{-7}% ocean topography, polar motion
- 10^{-8}% general relativity

Gravity data for the United States is readily available from the National Geodetic Survey, a United States federal agency, through their Gravity for the Redefinition of the American Vertical Datum (GRAV-D) program. This data is useful, for example, in determining where water flows because the exact gravity of an area can better predict where water flows than altitude changes can predict alone.

The reason that local gravity is not more important is because, while across the globe gravity may differ by about 1%, rarely are marksmen zeroing targets at one extreme (e.g., the Hudson Bay) and then shooting at another extreme (e.g., Mount Everest). Realistically, gravity only varies by 0.1% or

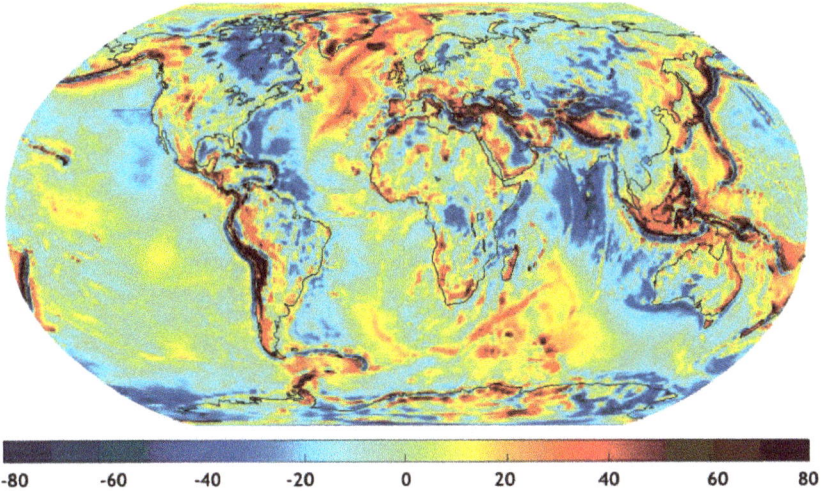

Image 106: This "gravity anomaly" map shows where Earth's gravity field **differs from a simplified model** that assumes the Earth is perfectly smooth and featureless. Areas colored yellow, orange, red, and brown are areas where the actual gravity field is larger than the featureless-Earth model predicts (e.g., the Himalayan Mountains in Central Asia). In contrast, the other shades indicate places where the gravity field is less (e.g., the area around Hudson Bay in Canada). This map was derived from NASA's Gravity Recovery and Climate Experiment (GRACE) Gravity Model 05. The color scale spans from −80 to +80 milligals (mGal), where one milligal is one thousandth of Earth's average surface gravity.

less at common shooting locations within a nation. Furthermore, marksmen typically zero and true in the area they are planning to shoot, eliminating even more of the difference.

10. Thermal Effects

Temperature is the measure of the average amount of heat in an area. Heat is how much energy atoms have. As atoms in a material increase or decrease their heat (i.e., energy), their physical properties change. For example, most materials expand as they heat up and contract as they cool down. These phenomena are called "thermal expansion" and "thermal contraction" respectively. Another change is in elasticity: as materials heat up, they usually stretch and deform more easily. In contrast, as materials cool down, they become more rigid. This phenomenon is called "thermoelasticity."

Thermal expansion is a significant problem (i.e., not de minimis) when an object contains multiple materials, such as a steel barrel attached to a stock (i.e., not freefloating), or wood in general (wood is composed of cellulose, lignin, and hemicellulose, and contains water). The effect of thermal expansion on rifle components has been well-documented. There was a famous instance of thermal expansion causing significant problems in 2015, in which the U.S. Government sued the company L-3 (EOTech's former parent company) for civil fraud. The U.S. accused L-3 of covering up defects in their rifle sights that the company knew about as early as 2006. One such defect was "thermal drift," which shifted the aiming points of the sights by as much as 30 cm (12 in) at 274 m (300 yd) as the sights heated or cooled.

It is exactly because of how well-known thermal expansion is, that manufacturers put in great effort to eliminate its negative effects. For example, almost all extreme long-range rifles do not use wood, and many favor metal chassis (a.k.a., stocks). These rifles also almost all have free floating barrels that prevent the barrel from touching the forestock.

The **de minimis effects are those that occur in single-material components**. For example, thermal expansion has long been suspected of causing issues in barrels. The hypothesis is that changing the physical properties of the barrel's metal alters how bullets travel through the bore and thereby affects a bullet's trajectory. This is borne out in the attention given to "cold-bore" shots, which is the first shot taken with a rifle. The classic thinking is that the first shot is especially inaccurate because the rifle barrel is the coldest, and therefore the smallest and most rigid, it will ever be during a shooting session.

However, there is no compelling evidence that the "cold-bore" inaccuracy exists. Of course the theory makes sense, and the thermal expansion does exist; rather, the lack of evidence points to the fact that the effect is so small as to be immeasurable. The lack of a measurable "cold-bore" effect could be the case for a few reasons.

First, **a single-metal object does not twist** due to a uniform temperature change. That is, if a free-floating barrel points one way, it always points that way, no matter the temperature. Similarly, a gas-powered semiautomatic usually has a gas tube made of the same material as the barrel, so it expands and contracts at the same rate. A metal barrel could deform if heating were unevenly distributed, but because the thermal conductivity of steel is approximately 2000 times that of air, heat quickly transfers to other parts of the rifle to evenly heat it.

Second, **the expansion itself is very small**. Carbon steel expands about 0.12% per 100°C (180°F). However, even a 100°C (180°F) change is a significant increase in temperature and rarely occurs in a barrel without

Thermal Expansion

continuous firing. If a 66 cm (26 in) barrel were heated up by 100°C (180°F), it would increase by a length of about 0.8 mm (0.03 in). And if the bore were 1 cm (0.4 in) in diameter, it would increase by 0.012 mm (0.0005 in). Notably, bore tolerance in mass-market rifles is already around 0.024 mm (0.0009 in). Such a tolerance is acceptable and performs well because bullets are designed to expand into rifling grooves as they travel through a bore.

That said, some marksmen do report a lower muzzle velocity from just a 0.012 mm difference in bore diameter due to the increased volume inside the bore for the gas to fill. The bore volume for a 100°C (180°F) increase amounts to 0.36%, or ~⅕ cm^3 (0.01 in^3). However even in this case, even though the gas must expand more into more volume, it has more time to do so since the bullet would take longer to leave the bore. Therefore, any decrease in velocity would be smaller than the increase in bore volume. Plus, to account for any increase in volume due to thermal expansion, a marksman would have to record and input the barrel's temperature as it actively changed, making any input by definition momentary and inaccurate.

Related to thermal expansion, some professionals claim that **heating a barrel during shooting relieves preexisting stresses** in the metal that are induced in the manufacturing process. These internal stresses do exist, but are already relieved by any reputable manufacturer by tempering the steel. Therefore, stress relief is a theoretical but unlikely possible source of heat-based inaccuracy for the first few shooting sessions with a new barrel.

Thermoelasticity is the relationship between the elastic properties of a material and its temperature. As materials heat up, materials stretch and deform more easily. The change in elasticity of steel is small at normal temperatures. Young's modulus (E) is a measurement that indicates how easily a material can stretch and deform. For steel, the Young's modulus decreases by roughly 2.16% for every 100°C (180°F) increase in temperature, within the range of 0°C to 300°C (32°F to 572°F). Even a 100°C (180°F) change is a significant increase in temperature, and firing a single bullet through a long-range rifle would not heat a barrel nearly that much. Therefore, there is no significant change to the Young's modulus from firing a single bullet through a long-range rifle.

A change in elasticity combined with a change in size could in theory affect a barrel's harmonics. "Barrel harmonics" is how the barrel shifts and moves as the shockwave of energy propagates through it from back to front. However, there is no reason to believe that a hot barrel would be significantly different from the harmonics of a cold one. Where harmonics do exist, they are very repeatable from one shot to the next, meaning they are accounted for during the zeroing process.

To cool down a rifle, a marksman can wait. But they can also slowly pour water down the barrel. As long as the barrel is below the tempering temperature of steel (up to approximately 150°C (302°F)), the micro-structure of the steel would remain unaffected no matter how quickly the steel cools.

11. Magnus Force

The Magnus force, or Magnus effect, occurs when fluid flows across an object **perpendicular to the spinning object's spin-axis**. For example, a crosswind blowing past a bullet perpendicular to its spin-axis generates a Magnus force. In contrast, a headwind or tailwind both align with the axis of a bullet's rotation, so do not induce a Magnus force.

The Magnus force is caused by the different forces of resistance on either side of the rotating body as a fluid, in this case air, passes over it. Fluid experiences **more resistance as it moves faster over a surface**. Therefore, if air blows into a rotating object or an object moves through static air, the air experiences more resistance on the side that is rotating towards the airflow's origin than that side that is rotating away from the airflow's origin. (Whether the object, the air, or both are moving is irrelevant; all that matters is that there is a speed differential between the two.)

More resistance causes slower movement. Slower movement causes more pressure due to Bernoulli's principle. Therefore, when an object rotates perpendicular to an incoming airflow, the object experiences a force that is 90 degrees from the air's direction of flow.

For example, for a right-twist bullet (i.e., spinning clockwise when viewed from the rear) in a right-to-left crosswind, there is a downward Magnus force. A left-twist bullet in the same crosswind experiences an upward force.

Magnus Force for Left-to-Right Crosswind
(Flip the red arrows for Right-to-Left Crosswind)

Right-Twist Bullet   Left-Twist Bullet

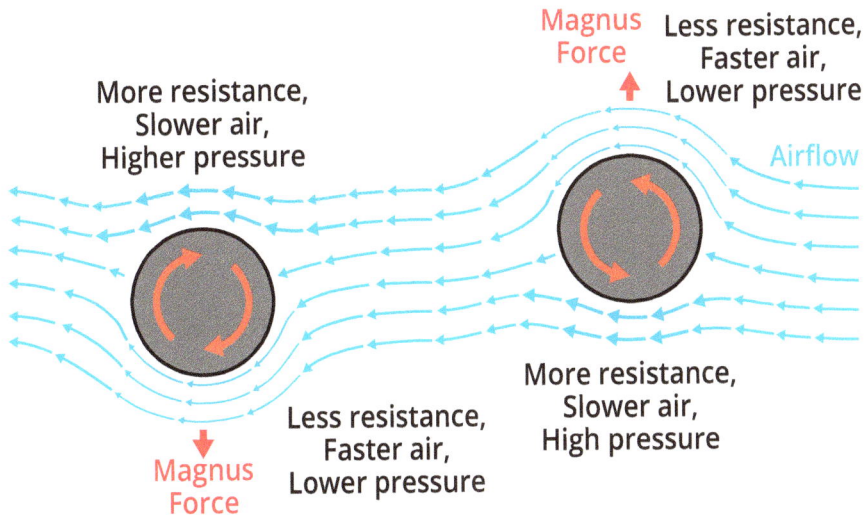

Magnus Force — Less resistance, Faster air, Lower pressure

More resistance, Slower air, Higher pressure

Airflow

Less resistance, Faster air, Lower pressure

Magnus Force

More resistance, Slower air, High pressure

Magnus Force

Image 109: The Magnus force occurs when a **spinning object** moves through and spins perpendicular to a **moving fluid**. This image is of the back of a bullet, where the airflow is a crosswind. The surface of the bullet that is spinning with the air causes less resistance than the surface spinning against the direction of the air. More resistance slows the air down more. Slower air causes more pressure according to Bernoulli's principle. Therefore, spinning in a fluid creates a force that is 90 degrees perpendicular to the direction the fluid is flowing. For bullets, this effect is negligible.

Reversing the direction of the crosswind also flips the direction of the Magnus force. In this way, **the Magnus force works in the opposite direction of aerodynamic jump**.

The magnitude of the Magnus force is determined by many factors. Because a major factor of resistance is friction, bullets that have rougher surfaces experience more Magnus force. The second factor is surface area for the same reason: a larger surface area has more friction than a smaller area. Third is the difference in relative speed that air experiences when blowing over either side of the bullet. The higher the relative speed differential (e.g., the faster the crosswind), the greater the Magnus force.

In practical terms, **the Magnus force on a bullet is insignificant** for a few reasons. First, the Magnus force depends on the surface of the bullet gripping the air. At supersonic speeds, the surface of the bullet (including the rifling marks) is covered by a boundary layer of air, reducing direct contact with the sideways airflow. Second, as the bullet travels in an airflow, it weathervanes into the direction of incoming air, progressively reducing the amount of air that passes over the bullet perpendicular to its spin-axis.

It is purported that the Magnus force can be measured in theory whenever winds exceed 15 km/h; however, since the force works directly opposite of a stronger force that scales in the same way (i.e., aerodynamic jump), there is no good way to measure it separately.

That said, the Magnus force can significantly affect a bullet's stability as it spins. Due to manufacturing imperfections, the bullet's spin-axis does not perfectly align with its center-of-mass. Additionally, as the bullet moves through air pockets with slightly different densities, its spin-axis also misaligns with the center-of-pressure. These factors cause the bullet to wobble in the air, resulting in complex Magnus-force effects. This destabilizing influence makes theoretical ballistic coefficients less accurate compared to those obtained from empirical measurements. However, these complex stability issues are not something that an individual marksman can effectively account for.

12. Coning and Nutation

In addition to one large precession (See Spin Stabilization (Precession), Pg. 74.), bullets also have secondary and simultaneous secondary rotations, called nutations. (The combined pattern of precession and nutations are sometimes referred to as "epicyclic swerve.") (See Image 111, Pg. 119.)

Nutation appears as a wobbling motion of the bullet's spin-axis around its precession-axis during flight. (See Image 110, Pg. 119.) Because nutations are **secondary rotations**, when they are averaged together, the precession is unaffected. Therefore, nutations are impractical to measure individually. And because they can be averaged together without affecting the precession, they have a nominal net effect on a bullet's trajectory.

13. Additional Effects

There are a few other forces that are ignored when calculating bullet trajectory because they are simply too small to detect or are only theoretical. For example, all texts assume that bullets are perfectly rigid to make calculations

Nutations

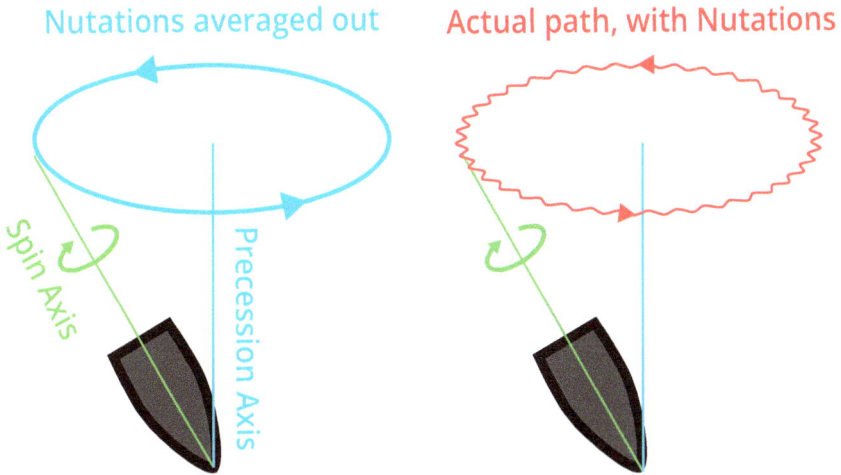

Image 110: Spin-stabilized objects actually have **two methods of rotation**. The first is the large circle that the spin-axis follows, called the precession. However, along the precession circle, the spin-axis also travels in many more smaller, **secondary circles, called nutations**. Nutations can vary greatly in size. Gyroscopes often have very small nutations as they precess in a circle. In contrast, bullets can have nutations that are very large in proportion to their primary circle of precession. Dynamic stability can be compromised if either the precession or the nutation become unstable. (See Image 96, Pg. 98.)

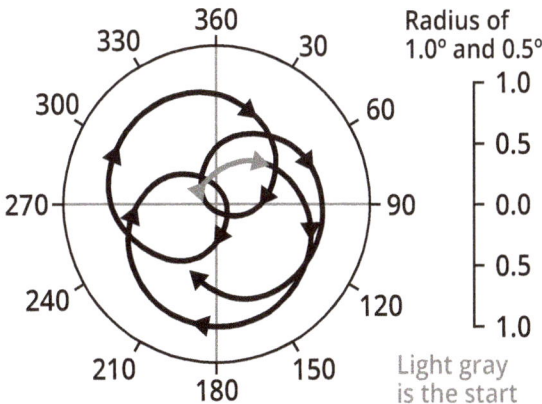

Image 111: This is a two-dimensional view of where an actual bullet's spin-axis pointed over time as the bullet traveled forward a set distance. Although the spin-axis followed a generally circular path, there were many smaller, secondary rotations (i.e., nutations) that complicated the curve.

simple; however, every material has its own elastic properties. In theory, bullets flex as they fly, affecting their ballistic properties ever so slightly.

Different material properties even apply to the air itself. While air density effects are considered for air resistance, different air density also affects other forces. For example, the Magnus force depends on the viscosity of the air. Also, as the bullet flies, the center-of-pressure shifts along the bullet, and how that happens is also influenced by the properties of the air. For another example, all objects in a fluid experience buoyancy forces, which are also affected by the properties of the air.

Another insignificant approximation is when continuous paths are entered into formulas as points. For example, when calculating the Coriolis effect, only the initial and final coordinates are used in the equation. However, as the bullet curves, it changes latitude, and a new instant Coriolis force would have to be calculated for complete accuracy. In theory, a true Coriolis force would require calculus to calculate; however, using just one coordinate is accurate enough for the purposes of long-range shooting.

The state of the rifle itself changes between different rounds. A common example is powder combustion residue, which is left over from each shot. Also, each shot degrades the chamber to some microscopic effect, which necessarily alters a bullet's path to an insignificant amount.

Some effects are theoretical, if not apocryphal. For example, the Poisson Effect supposes that the tip of the projectile above the direction of travel causes an air cushion to build up underneath it. Therefore, there is an increase of friction between this cushion and the bullet so that the bullet tends to spin off the cushion and move sideways.

In the future, scientists, engineers, and programmers may eventually not only develop computer programs that can account for these effects, but also build a robot that can take advantage of that program. However, as it stands now, **humans can accurately shoot for kilometers and miles while only using the knowledge carefully described in previous chapters.**

Best of luck shooting!

Appendices

Appendices

The shepherd drives the wolf from the sheep's throat, for which the sheep thanks the shepherd as his liberator, while the wolf denounces him for the same act as the destroyer of liberty.

Plainly, the sheep and the wolf are not agreed upon a definition of liberty.
—Abraham Lincoln, 16th commander-in-chief of the United States Army

14. Glossary

Aerodynamic Jump	The minor vertical shift in a bullet's trajectory caused by a crosswind rotating a bullet sideways.
Airflow	The movement of air across and over a bullet or object, influencing drag and flight stability.
Ambushing	A shooting technique where the marksman holds the sights still and fires the moment a moving target crosses the point-of-aim.
Angle-of-Yaw	The angle between a projectile's longitudinal axis and its actual flight path.
Angular Momentum	Angular momentum is momentum going in a circle. It's conserved, which means the total amount of momentum must be the same unless acted on by a force.
Average Velocity	The mean speed of a projectile over a specified distance.
Barrel Mirage	Heat waves rising from a warm barrel that create a visual distortion, potentially affecting the marksman's aim.
Barrel Twist	The rate at which a rifle's rifling spins a bullet, typically measured in centimeters or inches per turn.
Berm	An elevated earthen backstop designed to safely contain bullets and reduce ricochets on a shooting range.
Blasting Cap	A small, sensitive explosive component used to initiate a larger explosive charge.
Bullet Spin	The rotational motion imparted on a projectile by rifling to enhance its stability in flight.
Bullet Trace	The visible disturbance due to pockets of different air pressure left in a bullet's path, used by marksmen to spot impacts.
Bullet Wake	The turbulent airflow behind a projectile that influences drag and downrange stability.
Buoyancy Forces	The minor upward lift on a projectile caused by density differences in the surrounding air.
Chronograph	A device that measures a projectile's velocity by detecting its passage across multiple sensors.
Clip-Ons	Auxiliary optic or imaging attachments that mount directly onto a rifle's primary scope for specialized viewing.

Cone Cells	Photoreceptor cells in the human eye responsible for high-acuity color vision critical for precise aiming.
Coning	The slight circular or oscillatory motion of a spinning bullet's nose around its primary axis of flight. The spin-axis forms a cone shape as it rotates around the mean center-of-mass.
Coriolis Effects	The apparent drift of a bullet's trajectory due to Earth's rotation.
De Minimis	A term indicating that an effect or amount is so minor as to be considered negligible.
Detonating (Det) Cord	A flexible explosive line that rapidly transmits a detonation wave between connected charges.
Drag Coefficient	A dimensionless value indicating how smoothly or turbulently a projectile passes through air.
Dynamic Instability	The tendency of a spinning projectile to not align the spin-axis to the applied force. A dynamically unstable bullet does not align its spin-axis to the incoming airflow.
Eotvos Effect	A change in the amount of centrifugal force an object experiences when it is fired East or West.
Epicyclic Swerve	The small, cyclic oscillation of a spinning bullet's nose around its primary flight path. See nutation.
E-type NATO Target	A standard, life-size silhouette target commonly used in military training to simulate an upright human figure.
Fragmentation Bullets	Projectiles engineered to break apart on impact, creating multiple wound channels for increased lethality.
Free Floating Barrel	A rifle barrel design that ensures no contact with the stock or fore-end, reducing external pressures for improved accuracy.
Frost Trail	A transient condensation trail formed behind a projectile under specific cold and humid atmospheric conditions.
Gaskets	Sealing components used in firearms or explosive devices to prevent unwanted gas or fluid leakage.
Gimbal Stabilizer	A pivoting support mechanism that keeps mounted equipment (e.g., cameras or scopes) steady despite external movement.
Glint	A noticeable flash of light reflecting off a surface.
Grains	A unit of weight (1/7000th of a pound) commonly used to measure bullets and powder charges in ammunition.
Grooves	The helical channels cut into a barrel's rifling that grip the bullet and impart stabilizing spin.
Hashmark	The reference lines or markings on a scope reticle used for precise aiming, windage, or elevation adjustments.
Haunch	The upper part of a deer's thigh.
Headwinds	Winds blowing directly against a projectile's flight path, decreasing its velocity and requiring additional elevation compensation.
Helical Spinning	The corkscrew-like motion of a bullet traveling through the air due to its rifling-induced rotation.
Hollow-Point Bullets	Projectiles with a concavity in the nose designed to expand on impact for increased energy transfer and wound channel size.
Housings	Protective or structural enclosures (such as scope bodies) that safeguard internal components from environmental damage.
Incoming Airflow	The volume and direction of air meeting a projectile in flight, influencing its drag and overall trajectory.

Jerk	To pull a trigger too fast so that the rifle shakes. Also, the rate of change of acceleration (third derivative of velocity), often felt as a sudden jolt in recoil dynamics.
Lateral Throwoff	A sideways deviation in bullet flight caused by uneven barrel harmonics or slight imbalances in bullet spin.
Laze	To use a laser rangefinder to measure target distance, enabling precise ballistic calculations.
Lead	The forward aim point applied to track and hit a moving target, compensating for its direction and speed.
Loadouts	Configurations of ammunition and equipment that a marksman carries, tailored to specific mission or shooting requirements.
Mach	A measure of speed relative to the speed of sound (Mach 1), used to describe supersonic, transonic, or subsonic projectile travel.
Magnus Force	A lift force acting on spinning projectiles due to asymmetric airflow around their rotating surfaces.
Mean Center-of-Gravity	The averaged or effective point where the center-of-gravity for an object would be for a moving object if the mass at all locations were averaged together.
Mil	To measure angular distance with a reticle.
Milligals	A unit of acceleration (1/1000 of a gal) used in precise gravitational measurements relevant to ballistic calculations.
Movement Value	A fraction by which the actual movement speed is multiplied to attain the equivalent sideways movement speed, perpendicular to the bullet's trajectory.
Nutation	The small, conical wobble of a spinning bullet's nose around its circle of precession.
Over-stabilization	A myth that a projectile's spin can be so high that it resists aerodynamic forces, potentially compromising trajectory and terminal performance.
Overturning Force	An aerodynamic force acting on a spinning projectile's nose, attempting to tilt it away from its flight path. Most commonly, this is the lift force.
Parallax	An optical effect where the reticle and target appear to move relative to each other if the marksman's eye position shifts.
Pitch	The angular rotation of a projectile or firearm around its transverse axis, influencing elevation and point of impact.
Precession	A principle of gyroscopes (e.g., a spinning bullet) that states when a force is applied to a spinning object, the force occurs in part 90 degrees later in the direction of rotation.
Rifling	The helical grooves inside a firearm's barrel that impart spin to a bullet, enhancing its stability in flight.
Right-Twist	A rifling pattern in a barrel that rotates clockwise away from the marksman. A projectile spins clockwise when viewed from the back as it travels through the barrel.
Rod Cells	Photoreceptor cells in the human eye responsible for low-light and peripheral vision, aiding target detection in dim conditions.
Spall	Fragmented material that flakes or breaks off a surface (e.g., armor or bullet) upon impact or extreme stress.
Spider Strap	A multi-armed securing system used to stabilize equipment or a person. It is named as the different arms loosely resemble a spider's web.

Spin Drift	A phenomena whereby the precession of a spinning bullet transfers gravitational energy into a lateral force that pushes a bullet to the side of its twist. A right-twisted bullet drifts to the right and a left-twisted bullet to the left.
Spin-Stabilized	Describes a projectile that maintains stable flight primarily through precession.
Spotter	An observer who assists a marksman by estimating range, calling windage, and correcting shots for more precise impacts. The spotter is usually more experienced than their paired marksman.
Spotting Scope	A high-powered optic used to observe targets, bullet trace, and impacts at extended distances.
Static Stability	A gyroscope's (e.g., a bullet's) ability to resist a change in orientation and maintain its orientation in flight despite the forces being applied to it.
Supine Position	A shooting stance where the marksman lies on their back with the firearm angled forward toward the target.
Tailwinds	Winds blowing in the same direction as a projectile's flight, effectively increasing its speed and flattening its trajectory.
Thermal Expansion	The tendency of materials, including firearm barrels, to expand when heated, potentially affecting accuracy.
Trace	The mirage that follows a supersonic bullet as it changes the air density along its trajectory as it travels.
Tracers	Projectiles with an added incendiary compound that burns brightly, helping marksmen observe the bullet's path.
Tracking	Consistently adjusting sights to follow a moving target or observe bullet impact.
Tractability	The ability of a spin-stabilized bullet to reorient its spin-axis to the direction of incoming airflow.
Transonic Range	The range of speeds where some of the air surrounding a bullet travels at supersonic speeds while other air travels at subsonic speeds.
Truing	The process of adjusting a rifle's ballistic data or scope settings based on observed impact points at various distances.
Truing Bar	A rigid reference rod or device used to verify that an actual trajectory matches a ballistic calculator's predicted trajectory.
Tumbling	The end-over-end flipping of a projectile in flight.
Twist Rate	The distance in a barrel (in inches) required for the rifling to complete one full rotation of a bullet.
Vapor Trail	The visible condensation path (i.e., a cloud) behind a fast-moving bullet that is created under specific atmospheric conditions.
Wave Drag	The aerodynamic resistance created by shock waves forming around a supersonic projectile.
Weathervaning	A projectile's natural tendency to turn its nose into the oncoming airflow, reducing drag and improving stability.
Windage Hold	The horizontal sight adjustment (measured in angular units) applied to compensate for crosswind drift.
Wobble	A minor oscillation of a spinning bullet's nose that dampens over distance, affecting flight precision.
Yaw	The angular deviation between a bullet's longitudinal axis and its actual trajectory.

Yaw-of-Repose	A small, steady angular offset adopted by a spinning bullet due to aerodynamic and gyroscopic forces.
Young's Modulus	A measure of a material's stiffness, indicating its resistance to elastic deformation under stress (such as in barrel flex).

15. Credits

The explicit and implied contents herein do not imply or constitute endorsement by the U.S. DOD or any of its branches.

Front Cover: U.S.M.C. CPL Manuel A. Estrada
Back Cover 1: Louisiana Army N.G. SSG Greg Stevens
Back Cover 2: U.S. Army SPC Uriel Ramirez
Back Cover 3: U.S. Army SGT Patrik Orcutt
Back Cover 4: U.S.M.C. CPL Amelia Kang
TOC Image 1: U.S. Army SSG Austin Berner
TOC Image 2: U.S.M.C SGT Esdras Ruano
TOC Image 3: U.S.M.C LCPL Juan Carpanzano
CT Image 1: U.S.M.C CPL Brandon Salas
CT Image 2: U.S.M.C CPL Khalil Ross
CT Image 3: U.S. Army SPC Samuel Hyer
CT Image 4: U.S. Army SGT Patrik Orcutt
Image 1: U.S.M.C PFC Eric Keenan
Image 5: Scott Bauer
Image 8: U.S. Army 1LT Benjamin Haulenbeek
Image 9: Anne Barth
Image 10: U.S. Army 1LT Benjamin Haulenbeek
Image 11: U.S. Army Justin Connaher
Image 12: U.S. Army SGT Patrik Orcutt
Image 13: U.S.M.C CPL Kirstin Merrimarahajara
Image 14: U.S.M.C LCPL Paley Fenner
Image 15: U.S.M.C CPL Ricky Gomez
Image 16: U.S. Army N.G. SGT Remi Milslagle
Image 17: U.S. Navy Seaman Christopher Williamson
Image 18: U.S. Army SGT Patrik Orcutt
Image 19: U.S. Army K. Kassens
Image 21: Warren LeMay
Image 23: U.S.M.C CPL Dean Gurule
Image 24: U.S. Army K. Kassens
Image 25: U.S. Army SSG Russell Klika
Image 26: U.S. Army SGT Connor Mendez
Image 27: U.S. Army SPC Taylor Shaffer
Image 28: U.S. Army SPC Taylor Shaffer
Image 29: U.S. Army SSG Sidney Sale
Image 30: U.S. Air Force Justin Connaher
Image 31: U.S. Army SPC Zachary Bouvier
Image 32: U.S.M.C SSGT Jestin Costa
Image 33: U.S. Army K. Kassens
Image 34: U.S. Army SPC Taylor Shaffer
Image 35: U.S. Army SSG William Waller
Image 36: U.S. Army SGT Patrik Orcutt
Image 37: U.S.M.C SGT Tia Nagle
Image 38: U.S. Army K. Kassens
Image 39: U.S.M.C CPL Matthew A. Callaha
Image 40: U.S. Army SGT Connor Mendez
Image 41: U.S.M.C SGT Tia Nagle
Image 43: U.S.M.C SGT Tia Nagle
Image 44: U.S.M.C SGT Tia Nagle
Image 45: U.S. Army PVT Austin Anyzeski
Image 46: U.S. Army SPC Caroline Schofer
Image 47: U.S.M.C CPL Sean Potter
Image 48: U.S. Air Force SRA Brennen Lege
Image 49: Courtesy photo
Image 50: Courtesy photo
Image 51: U.S. Air Force SSGT Trevor McBride
Image 52: U.S.M.C CPL Israel Chincio
Image 53: U.S.M.C CPL Diana Jimenez
Image 54: U.S. Army SPC Michael J MacLeod
Image 55: U.S.M.C SGT Danny Gonzalez
Image 56: U.S.M.C SGT Isaac Ibarra

Image 57: U.S. Army SPC Kimberly Gonzalez
Image 58: JIP
Image 59: Kh1604
Image 60: U.S. Army PFC Lloyd Justine Villanueva
Image 61: U.S. Army SGT Karen Sampson
Image 62: U.S.M.C CPL Manuel A. Estrada
Image 63: U.S.M.C CPL Manuel A. Estrada
Image 65: U.S. Air Force CPT Kippun Sumner
Image 66: U.S. Army SGT Steven Lewis
Image 67: U.S. Army SSG Malcolm Cohens-Ashley
Image 69: U.S.M.C CPL Aaron Patterson
Image 70: U.S.M.C CPL Demetrius Morgan
Image 76: Albarubescens
Image 72: Pollyanna1919
Image 73: Vic2015
Image 87: Robert L. McCoy
Image 88: Bryan Litz
Image 89: Bryan Litz
Image 92: James A. Boatright and Gustavo F. Ruiz
Image 93: Hornady
Image 98: Bryan Litz
Image 99: Bryan Litz
Image 100: Bryan Litz
Image 101: Bryan Litz
Image 106: NASA and German Aerospace Center (DLR)
Image 107: Jacek Halicki
Image 108: Jan Polák
Image 111: James A. Boatright and Gustavo F. Ruiz
LRSU Front Cover: U.S. Air Force TSGT Gregory Brook
LRSU Back Cover 1: U.S. Army K. Kassens
LRSU Back Cover 2: U.S. Army CPT Thomas Cieslak
LRSU Back Cover 3: U.S.M.C. CPL Ricky S. Gomez
LRSU Back Cover 4: U.S. Army SSG Oscar Gollaz

Please leave a review. Thanks!

Positive reviews from awesome people like you help other people to benefit from the valuable instructions in this manual. Could you take 60 seconds to share your thoughts?

Thank you in advance for helping the community!

If you liked this book, consider buying, *Long Range Shooting.*

Or consider both books together, *Long Range Shooting United.*

www.ingramcontent.com/pod-product-compliance
Lightning Source LLC
Chambersburg PA
CBHW040902210326
41597CB00029B/4937